MW01030230

FM 23-6

U. S. RIFLE, CALIBER .30
M1917 ENFIELD
FIELD MANUAL

August 3, 1942

BY WAR DEPARTMENT

ISBN#978-1-940453-14-9
www.PeriscopeFilm.com

DISCLAIMER:

This manual is sold for historic research purposes only, as an entertainment. It contains obsolete information and is not intended to be used as part of an actual operation or maintenance training program. No book can substitute for proper training by an authorized instructor.

ISBN#978-1-940453-14-9
www.PeriscopeFilm.com

FM 23-6

BASIC FIELD MANUAL

U. S. RIFLE, CALIBER .30 M1917 (ENFIELD)

UNITED STATES
GOVERNMENT PRINTING OFFICE
WASHINGTON : 1942

For sale by the Superintendent of Documents, Washington, D. C.
Price 35 cents

WAR DEPARTMENT,
WASHINGTON, August 3, 1942.

FM 23-6, Basic Field Manual, U. S. Rifle, Caliber .30, M1917 (Enfield), is published for the information and guidance of all concerned.

[A. G. 062.11 (6-8-42).]

BY ORDER OF THE SECRETARY OF WAR:

G. C. MARSHALL,
Chief of Staff.

OFFICIAL:

J. A. ULIO,
Major General,
The Adjutant General.

DISTRIBUTION:

C and H (5); IBn 1 (10); IC (20).
(For explanation of symbols see FM 21-6.)

TABLE OF CONTENTS

BASIC FIELD MANUAL

U. S. RIFLE, CALIBER .30, M1917 (ENFIELD)

CHAPTER 1

MECHANICAL TRAINING

SECTION I

DESCRIPTION

■ 1. GENERAL.—The United States rifle, caliber .30, M1917 (fig. 1), is a breech-loading, magazine rifle of the bolt type. It is sometimes called the *Enfield rifle*.

■ 2. PRINCIPAL DIMENSIONS, WEIGHTS, AND MISCELLANEOUS DATA.

Weight, without bayonet	pounds	9. 187
Weight, with bayonet	do	10. 312
Length, without bayonet	inches	46. 3
Length, with bayonet	do	62. 3
Diameter of bore	do	. 30
Trigger pull, minimum	pounds	3. 0
Rifling:		
Number of grooves		5
Twist, uniform, left hand, one turn in	inches	10
Sight radius (distance from top of front sight to rear side of leaf, leaf raised)	do	31. 76
Sight radius (battle sight)	do	31. 69
Magazine capacity	rounds	6

■ 3. REAR SIGHT.—The rear sight of this rifle has no wind gage. The leaf contains a peep sight which moves vertically on a slide, and hence makes no correction for the drift of the bullet. The battle sight is of the peep type and is attached to the lower end of the leaf. It is raised to position for aiming when the leaf is laid, and is adjusted for a range of 400 yards.

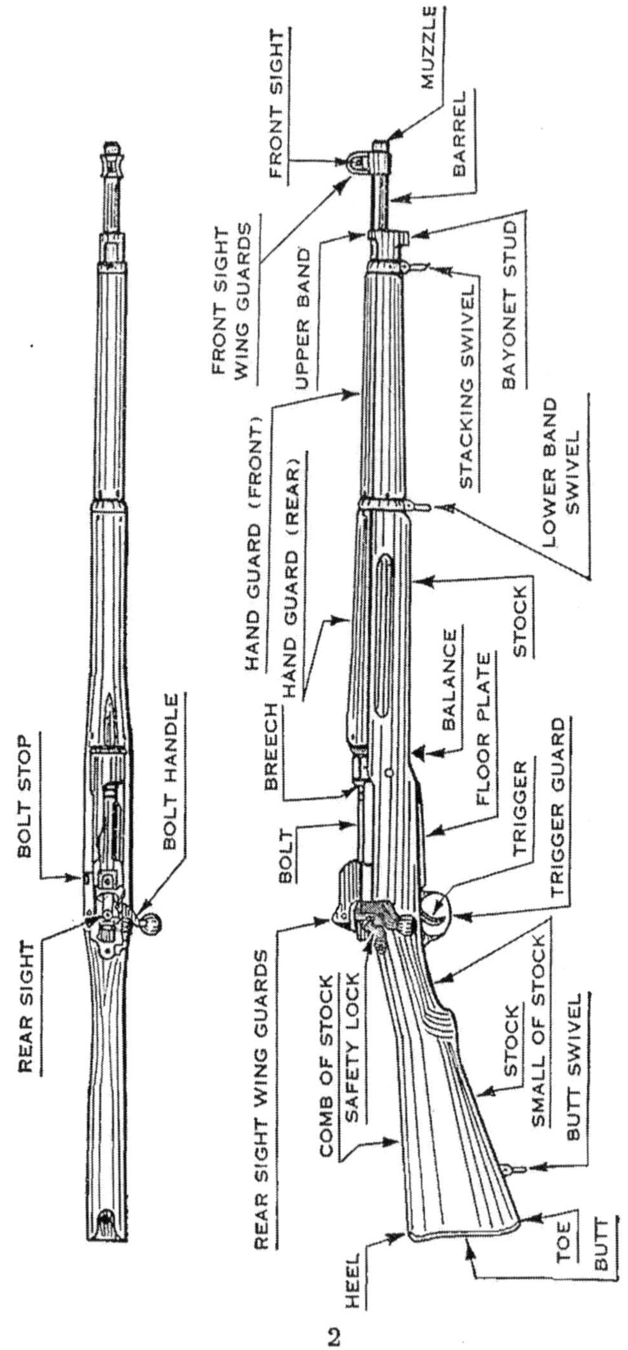

FIGURE 1.—U. S. rifle, caliber .30, M1917.

The leaf is graduated from 200 to 1,600 yards. It is graduated in multiples of 100 yards from 200 to 900 yards and in multiples of 50 yards from 900 to 1,600 yards. (See fig. 2.)

■ 4. FRONT SIGHT.—The front sight is protected by wing guards, and is adjusted laterally during assembly at the arsenal. It is locked in position, after adjustment, by upsetting part of the metal base into the lock seat with a punch.

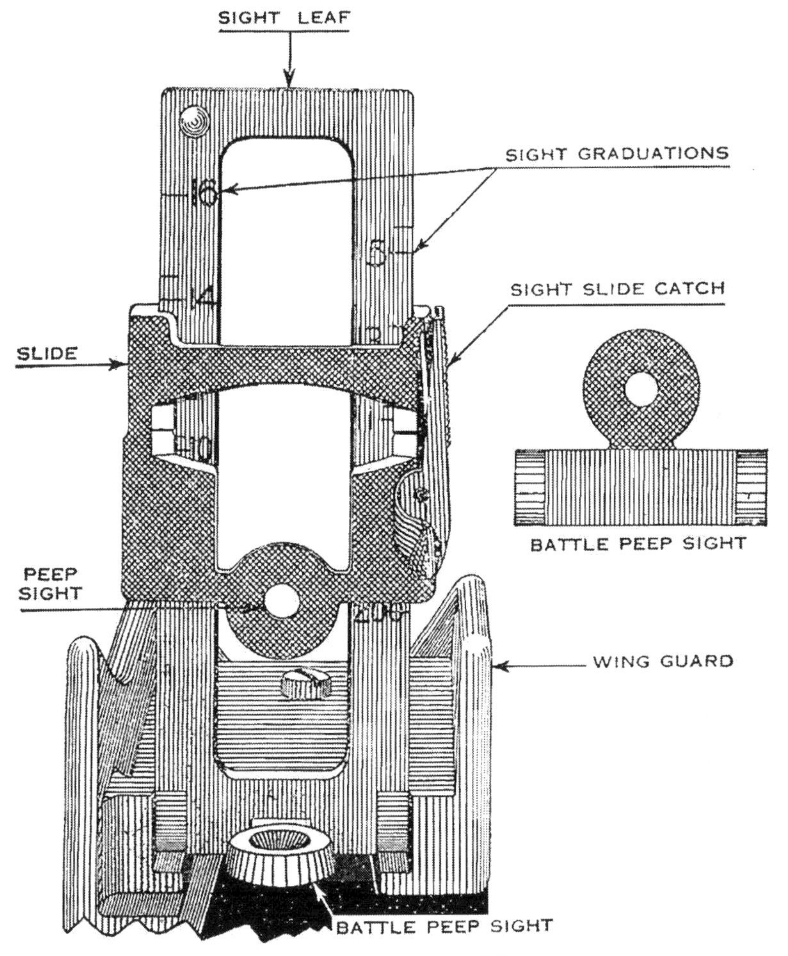

FIGURE 2.—Rear sight.

■ 5. RATE OF FIRE AND EFFECTIVE RANGE.—The maximum rate of accurate fire with this weapon depends upon the skill and the position of the operator and the visibility of the target. It varies with the individual's ability and practice; a well-trained soldier should be able to fire from 10 to 15 shots per minute. The effectiveness of rifle fire decreases as the range to the target increases. Its use at ranges greater than 600 yards is unusual.

■ 6. NOMENCLATURE AND REFERENCES.—*a. Nomenclature.*—The soldier should be familiar with the names of those parts of the rifle which are frequently referred to in drill and range practice (figs. 1 and 3).

b. References.—Safety precautions to be observed by troops are complete in this manual. Range officers, the officer in charge of firing, and the commander responsible for the location of ranges and conduct of firing should refer to AR 750–10 for the location of ranges and conduct of firing.

SECTION II

DISASSEMBLY AND ASSEMBLY

■ 7. PLACE IN TRAINING.—Disassembly and assembly should be taken up as soon as practicable after the soldier receives his rifle. In any event this training is completed before the individual does any firing with the rifle. Instruction in the care and cleaning of the rifle is also covered before it is fired.

■ 8. DISASSEMBLY.—*a. General.*—Disassembly of the rifle by the individual soldier is limited to that required for proper care and maintenance of the rifle. Further disassembly is done under the supervision of an officer or by ordnance personnel. Only the following parts of the rifle may be removed by the individual soldier for the purpose of cleaning:

(1) Floor plate and follower.
(2) Gun sling.
(3) Oiler and thong case.
(4) Bolt.

b. To remove and disassemble floor plate and follower.—Insert the bullet end of a cartridge through the hole in the floor plate and press down on the floor plate catch; at the same time draw the floor plate to the rear with the left thumb. This releases the floor plate which may then be

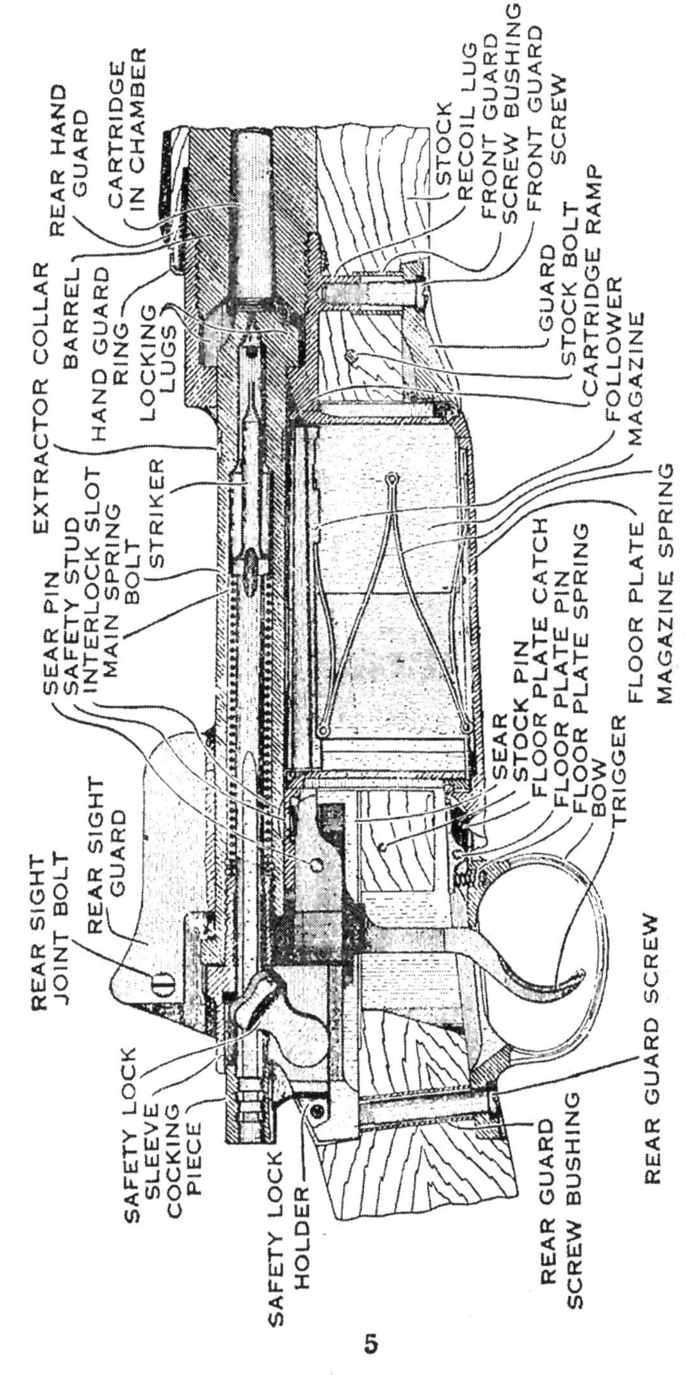

FIGURE 3.—Cross section of mechanism.

removed together with the magazine spring and follower. Raise the rear end of the magazine spring until it clears

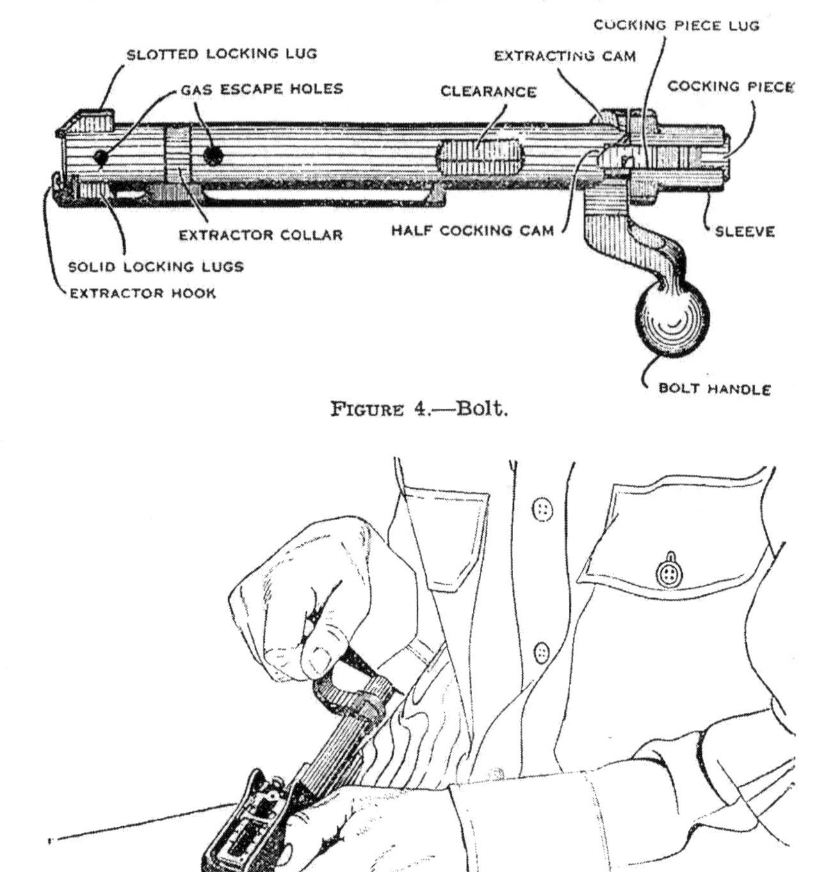

FIGURE 4.—Bolt.

FIGURE 5.—Removing bolt.

the spring stops on the floor plate and draw it out of its mortise; in the same manner separate the magazine spring from the follower. To assemble, proceed in reverse order.

c. To remove and disassemble bolt.—Place the butt of the rifle under the right armpit, and hold the stock firmly against the body with the right arm; with the left thumb pull the bolt stop to the left and at the same time raise the bolt handle and draw out the bolt to the rear. (See figs. 4 and 5.) Hook a loop of strong string on the dismounting hook on the cocking piece lug and, holding the bolt in the left hand and the string in the right, draw the cocking piece out until the lug clears the end of the bolt (see fig. 6).

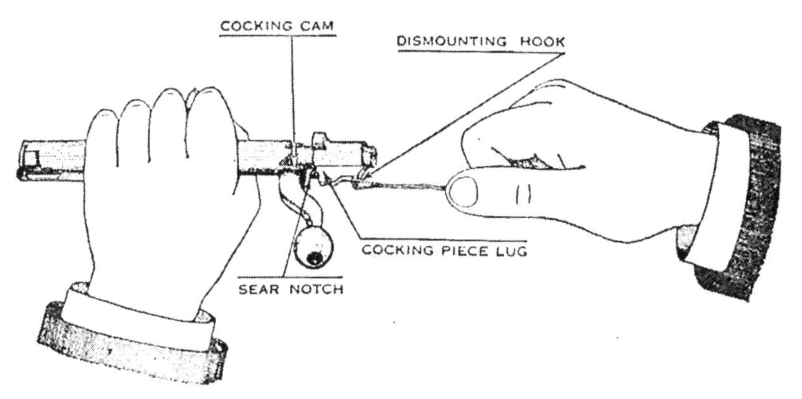

FIGURE 6.—Withdrawing cocking piece.

Then, by moving the right hand in a circular path counter-clockwise, unscrew the sleeve, and withdraw the sleeve, cocking piece mainspring, and striker from the bolt. Grasp the sleeve with the thumb and forefinger of the left hand; place the point of the striker against a hard surface; and force the sleeve downward, compressing the mainspring until the lug on the cocking piece clears the lug slot in the sleeve, as shown in figure 7. Then with the right hand rotate the cocking piece a quarter turn in either direction to disengage it from the striker, and draw it off to the rear. Relieve the spring from the stress slowly and remove it and the sleeve from the striker, being careful that the parts do not fly from the hand. Turn the extractor to the right so that it covers the gas-escape holes in the bolt; place the right thumb about midway between the extractor collar and the lower end of the extractor and push downward and forward to force the extractor off the bolt. (See fig. 8.)

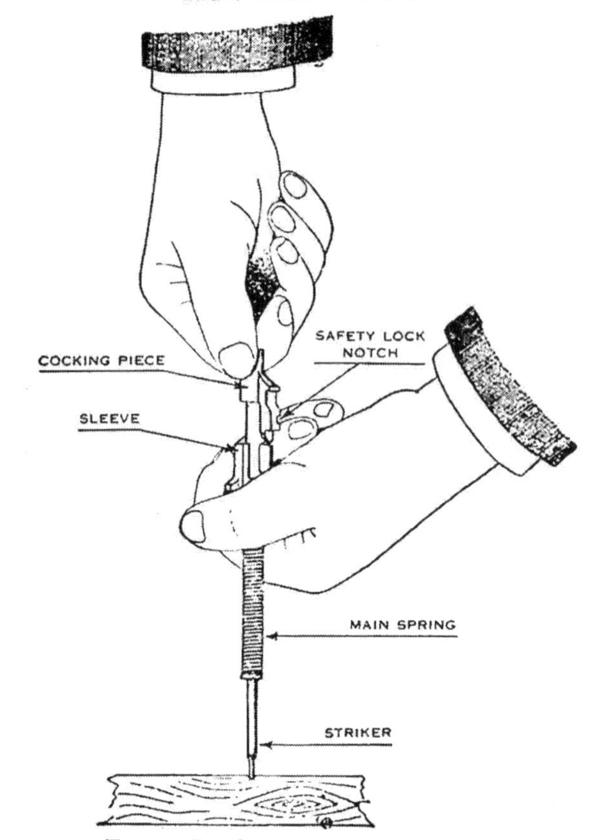

FIGURE 7.—Compressing mainspring.

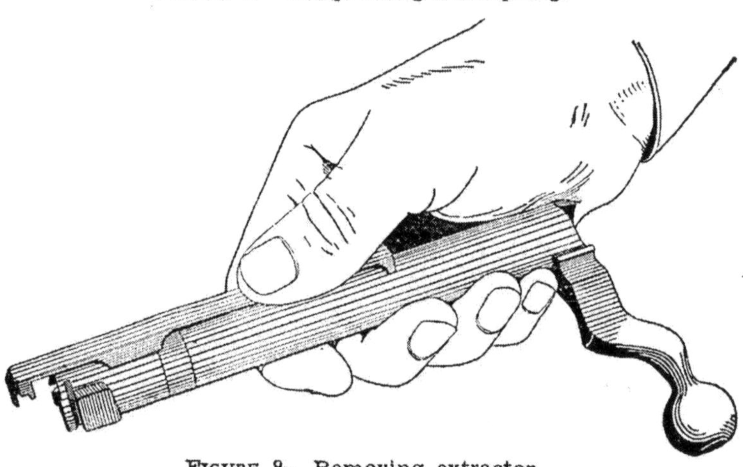

FIGURE 8.—Removing extractor.

■ 9. ASSEMBLY.—*To assemble and replace bolt mechanism.*—
a. Turn the extractor collar until its lug is on line with the gas-escape holes; insert the lug on the collar in the undercuts in the extractor by pushing the extractor to the rear until its tongue comes in contact with the face of the bolt; press the hook of the extractor against some rigid object until the tongue will slide over the end of the bolt. Turn the extractor so that it lies over the unslotted or solid locking lug. (See fig. 4.)

b. Slide the mainspring over the striker. Hold the point of the striker against a hard surface; place the sleeve against the end of the spring with the flat sides in its bore coincident with the flat sides on the striker; force the sleeve down on the striker, compressing the main spring. Holding the sleeve with the mainspring fully compressed, replace the cocking piece on the end of the striker, and lock it by a quarter turn so that its lug is alined with the lug slot in the sleeve. Then let the sleeve return to its position slowly under action of the spring. Grasp the bolt in the left hand and start the threads on the barrel of the sleeve into the threads in the end of the bolt. Hook the loop of string on the dismounting hook and, holding the ends of the string in the right hand, pull the cocking piece out and screw the sleeve home in the bolt by turning it clockwise. (See fig. 6.)

c. Hold the piece under the floor plate in the fingers of the left hand, the thumb extending over the left side of the receiver; take the bolt in the right hand with the cocking piece lug down; press the rear end of the follower down with the left thumb; push the bolt into the receiver; and lower the bolt handle.

SECTION III

CARE AND CLEANING

■ 10. IMPORTANCE.—The care and cleaning of the rifle is an important duty to be performed by all soldiers armed with this weapon, and the subject merits the serious consideration of all officers. Experience has shown that the majority of those rifles that become unserviceable do so through lack of intelligent and proper care and not from firing.

■ 11. LUBRICANTS, CLEANING MATERIALS, AND RUST PREVENTIVES.—The following are the only materials authorized and issued for cleaning these rifles. The use of unauthorized materials such as abrasives is forbidden.

Cleaner, rifle-bore.
Oil, lubricating, preservative, light.
Oil, lubricating, for aircraft instruments and machine guns.
Compound, rust-preventive, light.
Solvent, dry-cleaning.
Oil, linseed, raw.
Oil, neat's-foot.

a. Rifle-bore cleaner.—(1) Rifle-bore cleaner is issued for cleaning the bore of the rifle after firing. This material possesses rust-preventive properties and will provide temporary protection against rust after the bore has been cleaned with it. It is preferred, however, that the bore be dried immediately after cleaning and the metal coated lightly with light preservative lubricating oil.

(2) Rifle bore cleaner will freeze at temperatures below 32° F. If frozen, it must be thawed and shaken well before using. Closed containers should not be filled more than three-fourths full in freezing weather, as full containers will burst if the contents freeze.

b. Light preservative lubricating oil.—This oil has rust-preventive as well as lubricating properties but cannot be depended upon to provide protection from rust for long periods. It is used for the lubrication of all moving parts and for short-term protection against rust of all metal parts of the rifle. Its preservative action results partly from the oily film on the metal parts and partly from chemical combination of inhibitors in the oil with the metal. It will therefore protect the metal surfaces from rust even though no appreciable film of oil is present on the metal parts. When used on moving parts, however, it is necessary to maintain a thin film of oil to provide the necessary lubrication.

c. Lubricating oil for aircraft instruments and machine guns.—This oil may be used for lubricating the rifles when light preservative lubricating oil is not available. It is an extremely light oil which relies entirely upon maintenance of a film to protect metal surfaces from rusting. When it is used as a preservative, the metal parts must be inspected daily for rust, cleaned and again lightly coated with the oil.

d. Light rust-preventive compound.—This compound is issued for the protection of metal parts for long periods of time while the rifles are boxed and in storage. It is a sluggish liquid at about 80° F. and should be warmed before application. If heating facilities are not available, it can be brushed on to the parts at 80° F. or above.

e. Dry-cleaning solvent.—This is a noncorrosive petroleum solvent used for degreasing the rifles. It will remove grease, oil, or rust-preventive compound. It is generally applied with rag swabs to large parts and used as a bath for small parts. The surfaces must be thoroughly dried with clean rags immediately after removal of the solvent. Gloves should be worn by persons handling such parts after cleaning to avoid leaving finger marks, which are ordinarily acid and induce corrosion. Cleaning solvent will attack and discolor rubber.

f. Raw linseed oil.—This oil is used on wooden parts of the rifle to prevent drying and to preserve the stocks and hand guards.

g. Neat's-foot oil.—This is a pale yellow animal oil. It is used for the preservation of leather equipment such as gun slings.

■ 12. INSPECTION.—Unless otherwise ordered, rifle bores at inspections will be protected with a thin film of light preservative lubricating oil. If this is not available, lubricating oil for aircraft instruments and machine guns will be used. The mechanism of the rifles is lubricated in the same way. The oiler will be filled with light preservative lubricating oil. If the inspector wishes to examine the rifles minutely, he may order the oil to be removed from them. When this is done, unit commanders will see that the proper oil (as indicated above) is applied immediately after the inspection.

■ 13. Disassembly for Cleaning.—Only the following parts of the rifle will be removed by the soldier for the purpose of cleaning:

a. Floor plate and follower.

b. Gun sling.

c. Oiler and thong case.

d. Bolt.

■ 14. Care and Cleaning When No Firing Is Done.—*a.* This includes the ordinary care of the rifle to preserve its condition and appearance during the periods when no firing is being done. Rifles in the hands of troops should be inspected daily to insure proper condition of cleanliness. Schedules should allow time for cleaning rifles on each day the rifles are used in training.

b. Damp air and sweaty hands induce rust. The rifle should be cleaned and protected after every drill and after tours of guard duty. Special care should be taken when the rifles have been used on rainy days.

c. The bore of the rifle will always be cleaned by inserting a cleaning rod into the breech to avoid possible injury to the rifling at the muzzle. The bolt must be removed for this purpose. A barrack cleaning rod should be used. The thong and brush should be used only when the barrack cleaning rod is not available. To clean the bore, assemble a cloth patch to the cleaning rod and insert into the bore at the breech end. Move the cleaning rod and patch forward and backward several times through the bore and replace with a clean patch. Be sure a patch goes all the way through the bore before the direction is reversed. This will prevent the patch and rod from becoming stuck in the bore. Repeat until a patch comes out clean. This cleaning removes accumulations of dust, dirt, and thickened oil in the bore. After the bore has been thoroughly cleaned, saturate a patch with light preservative lubricating oil and push it through the bore to apply a light film of oil. When issue patches are not available, patches should be cut to approximately 2½ inches square to permit their passage through the bore without bending the cleaning rod.

d. The chamber of the rifle must be cleaned as thoroughly as the bore. A rough chamber may cause shells to stick.

Rub the metal surfaces of the rifle, including the bolt mechanism and magazine, with a dry cloth to remove moisture, perspiration, and dirt. Then wipe with a cloth which has been slightly oiled with light preservative lubricating oil. Apply a few drops of light preservative lubricating oil to all cams and working surfaces of the mechanism. To clean the outer surfaces of the rifle, wipe off dirt with a slightly oiled cloth and wipe dry with a soft, clean one, using a small cleaning brush or small stick to remove dirt from screwheads and crevices on the outside of the rifle.

e. After cleaning and protecting the rifle as described above, place it in the rifle rack without any covering and without a plug in the muzzle or bore. Muzzle covers, gun covers, rack covers, and plugs must not be used because they cause sweating and promote rust. When the barracks are being swept, rifle racks should be covered to protect the rifles from dust. These covers must be removed immediately after the rooms have been swept.

■ 15. Preparatory to Firing.—Before firing, take the following steps to insure efficient functioning of the rifle:

a. Dismount the main groups.

b. Clean the bore to remove dust, dirt, and thickened oil. Do not oil the chamber. Thoroughly clean and lightly oil all metal parts with light preservative lubricating oil. Be sure to apply a thin coating of this oil to all cams and working surfaces of the mechanism, including the bolt mechanism and magazine. Assemble the rifle and rub all outer surfaces with a lightly oiled rag to remove all dust. Then wipe off with a clean, dry rag.

■ 16. After Firing.—*a*. The bore and chamber of the rifle must be thoroughly cleaned not later than the evening of the day on which the rifle is fired. They should be cleaned in the same manner for the next 3 days. Use rifle-bore cleaner if available. If rifle-bore cleaner is not available, water may be used; warm water is good, but warm, soapy water is better. Hold the rifle muzzle down. Insert the cleaning rod assembled with a patch saturated with rifle-bore cleaner in the bore and move it forward and backward for about 1 minute. Be sure the patch goes all the way through the bore before the direction is reversed to prevent its becoming stuck in the bore. While the bore is wet a clean brush, if available, should be run all the way through and all the way back three or four times to remove any hardened particles in the bore. Remove the brush and run several patches saturated with cleaner or with water entirely through the bore, removing them from the muzzle end. Then wipe the cleaning rod dry and, using dry, clean patches, thoroughly swab the bore until it is perfectly dry. Be sure that the chamber is also dry and clean. Examine the bore and chamber carefully for cleanliness. If it is not free from all residue, repeat the cleaning process. When it is clean, saturate a patch with light preservative lubricating oil and push it through the bore so that a light film of the oil is deposited on the bore and chamber.

b. If the rifle is to be fired the next day, proceed as in paragraph 15. *If the rifle is not to be fired within the next few days, repeat the procedure outlined in "a" above for the next 3 days and complete cleaning as follows:*

(1) Deposits of primer salts in the chamber attract water and therefore cause rust. The salts can be removed only by use of rifle-bore cleaner, soapy water, or water alone. Saturate a cleaning patch with rifle-bore cleaner or with water and assemble to a small stick. Clean by twisting the patch-covered stick in the chamber in order to dissolve any primer fouling. Dry the chamber with successive dry patches. Inspect the chamber visually, then by inserting the little finger into the chamber and twisting it. If no discoloration shows on the finger, oil the chamber lightly with light preservative lubricating oil. Be sure that this oil is removed with a dry patch before the rifle is fired.

(2) Wipe the exterior of the rifle with a dry clean cloth to remove dampness, dirt, and perspiration. Wipe all metal surfaces with a cloth dampened with light preservative lubricating oil. Oil the stock and hand guard with raw linseed oil and, if sling is of leather, oil it with neat's-foot oil.

(3) The face of the bolt should be cleaned with a wet patch, dried and lightly oiled with light preservative lubricating oil.

■ 17. ON RANGE AND IN FIELD.—The rifle must be kept clean, free from dirt, and properly lubricated. To obtain its maximum efficiency, the following points must be observed:

a. Never fire a rifle with dust, dirt, mud, or snow in the bore.

b. Keep the chamber free from oil and dirt.

c. Never leave a patch, plug, or other obstruction in the chamber or bore. Failure to observe this precaution may result in serious injury if the rifle is fired.

d. Keep a light coating of light preservative lubricating oil on all metal parts. When the rifle gives indication of lack of lubrication and excessive friction, apply additional oil to the cams and working surfaces.

e. During range firing, a qualified man should be placed in charge of the cleaning of rifles at the cleaning racks or tables.

f. The oiler should be kept filled with light preservative lubricating oil.

g. When the barrack cleaning rod is not available, the contents of the oiler and thong case carried in the butt of the stock may be used to clean the rifle. In cleaning the bore by means of the thong, the brush or rag should be drawn from the breech toward the muzzle. The oiler should always be inserted into the stock so the leather-tipped cap will be next to the butt plate cap. This prevents noise in carrying the piece.

h. When the prescribed lubricants are not available, any clean light mineral oil such as engine oil may be used. For cleaning the bore and chamber, clean water may be used in place of rifle bore cleaner.

■ 18. PREPARATION FOR STORAGE.—*a.* Light preservative lubricating oil is the most suitable oil for short-term preservation of the mechanism of the rifle. It is effective for storage for periods of 2 to 6 weeks, depending on climatic conditions. However, rifles in short-term storage must be inspected every 4 or 5 days and the preservative film renewed if necessary. For longer periods of time, rifles should be protected with light rust-preventive compound.

b. Light rust-preventive compound is a semisolid material. It is efficient for preserving the polished surfaces, the bore, and the chamber for a period of 1 year or less, dependent on climatic and storage conditions.

c. The rifles should be cleaned and prepared for storage with particular care. The bore, all parts of the mechanism, and the exterior of the rifles should be thoroughly cleaned and then dried completely with rags. In damp climates, particular care must be taken to see that the rags are dry. After drying a metal part, the bare hands should not touch that part. All metal parts should then be coated with either light preservative lubricating oil or light rust-preventive compound, depending on the length of storage required. (See *a* and *b* above.) Application of the rust-preventive compound to the bore of the rifle is best done by dipping the cleaning brush in the compound and running it through the bore two or three times. The brush must be clean before use. Before placing the rifle in the packing chest see that the bolt is in its forward position and that the firing pin is released. Then, handling the rifle by the stock and hand guard only, it should be placed in the packing chest, the wooden supports at the butt and muzzle having previously been painted with rust-preventive compound. *Under no circumstances will a rifle be placed in storage contained in a cloth or other cover or with a plug in the bore.* Such articles collect moisture and cause the weapon to rust.

■ 19. CLEANING OF WEAPONS AS RECEIVED FROM STORAGE.—Weapons which have been stored in accordance with the previous paragraph will be coated with either light preservative lubricating oil or with light rust-preventive compound. Weapons received from ordnance storage will in general be coated with rust-preventive compound. Use dry-cleaning solvent to remove all traces of the compound or oil. Take particular care that all recesses in which springs or plungers operate are cleaned thoroughly. Failure to do this may cause malfunctioning at normal temperatures and will certainly do so when the rust-preventive compound congeals at low temperatures. After using the cleaning solvent, be sure it is completely removed from all parts by wiping with a dry cloth. Then follow the instructions in paragraph 14.

■ 20. CARE AND CLEANING UNDER UNUSUAL CONDITIONS.—*a. Cold climates.*—(1) In temperatures below freezing, it is necessary that the moving parts of the weapon be kept absolutely free from moisture. It has also been found that excess oil on the working parts will solidify to such an extent as to cause sluggish operation or complete failure.

(2) The metal parts of the weapon should be taken apart and completely cleaned with dry-cleaning solvent before use in temperatures below 0° F. The working surfaces of parts which show signs of wear may be lubricated by rubbing with a cloth slightly dampened with light preservative lubricating oil. At temperatures above 0° F., all metal surfaces of the weapon may be oiled lightly (after cleaning) by wiping with a slightly oiled cloth, using light preservative lubricating oil.

(3) When it is brought indoors, the weapon should first be allowed to come to room temperature. It should then be disassembled, wiped completely dry of the moisture which will have condensed on the cold metal surfaces, and thoroughly oiled with light preservative lubricating oil. If possible condensation should be avoided by providing a cold place in which to keep rifles when not in use. For example, a separate cold room with appropriate racks may be used or, when in the field, racks under proper cover may be set up outdoors.

(4) If the rifle has been fired, it should be thoroughly cleaned and oiled. The bore may be swabbed out with an oily patch, and when the weapon reaches room temperature thoroughly cleaned and oiled as prescribed in (3) above.

(5) Before firing, the weapon should be cleaned and oil removed as prescribed in (2) above. The bore and chamber should be entirely free of oil before firing.

b. Hot climates.—(1) *Tropical climates.*—(*a*) In tropical climates where temperature and humidity are high, or where salt air is present, and during rainy seasons, the weapon should be thoroughly inspected daily and kept lightly oiled when not in use. The groups should be dismounted at regular intervals and, if necessary, disassembled sufficiently to enable the drying and oiling of parts.

(*b*) Care should be exercised to see that unexposed parts and surfaces are kept clean and oiled.

(*c*) Light preservative lubricating oil should be used for lubrication.

(*d*) Wooden parts should also be inspected to see that swelling caused by moisture does not bind working parts. (If swelling has occurred, shave off *only enough* wood to relieve binding.) A light coat of raw linseed oil applied at intervals and well rubbed in with the heel of the hand will help to keep moisture out. Allow oil to soak in for a few hours and then wipe and polish the wood with a dry, clean rag. Care should be taken to see that linseed oil does not get into the mechanism or on metal parts as it will gum up when dry. Stock and hand guard should be dismounted when this oil is applied.

(2) *Hot, dry climates.*—(*a*) In hot, dry climates where sand and dust are likely to get into the mechanism and bore, the weapon should be wiped clean daily, or oftener if necessary. Groups should be dismounted and disassembled as far as necessary to facilitate thorough cleaning.

(*b*) When the weapon is being used under sandy conditions, all lubricant should be wiped from the weapon. This will prevent sand carried by the wind from sticking to the lubricant and forming an abrasive compound which will ruin the mechanism. Immediately upon leaving sandy terrain, the weapon must be relubricated with light preservative lubricating oil.

(*c*) In such climates, wooden parts are likely to dry out and shrink. A *light* application of raw linseed oil applied as in (1) (*d*) above will help to keep wood in condition.

(*d*) Perspiration from the hands is a contributing factor to rust because it contains acid. Therefore, metal parts should be wiped dry frequently.

(*e*) During sand or dust storms, breech and muzzle should be kept covered, if possible.

SECTION IV

FUNCTIONING

■ 21. OPERATION.—*a. To load.*—To load the magazine, raise the bolt handle, draw the bolt fully to the rear, and insert the cartridges from a clip or singly from the hand. To load from a clip, place either end of a loaded clip in the clip slot in the receiver and, with the fingers of the right hand under the rifle against the floor plate, and the base of the thumb on the powder space of the top cartridge near the clip, press the clip down into the magazine with a firm, steady pressure of the thumb until the top cartridge is caught by the right edge of the receiver (see fig. 9). The empty clip is removed with the right hand. After loading the maga-

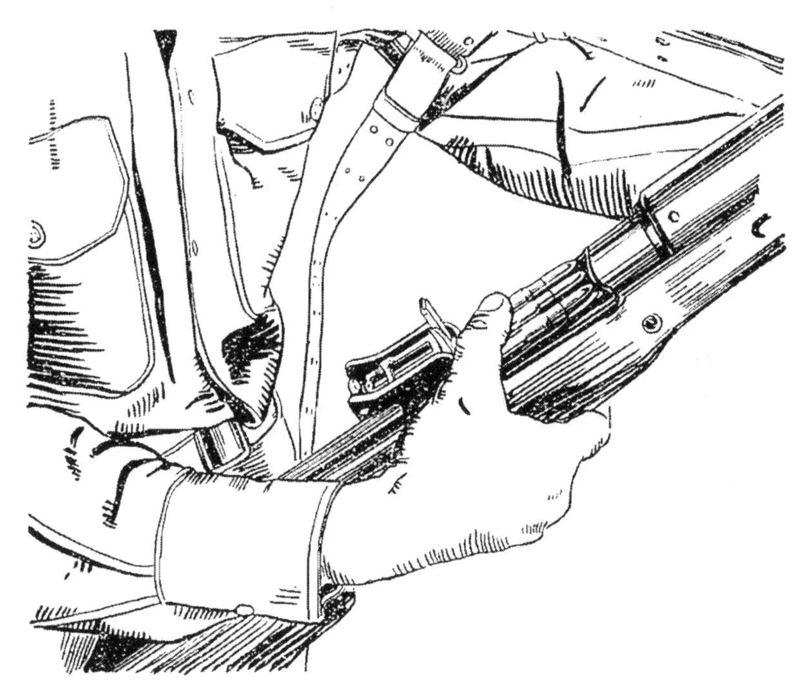

FIGURE 9.—Loading from clip.

zine, closing the bolt places a round in the chamber. The forward movement of the bolt pushes the top cartridge into the chamber. A single cartridge may be inserted by hand directly into the chamber. When this is done it is not necessary to open the bolt all the way to the rear.

b. Extraction.—The extractor begins to extract an empty case from the chamber during the rotation of the bolt when the bolt handle is raised, and completes the extraction as the bolt is drawn to the rear.

c. Ejection.—When the bolt is drawn fully to the rear, the head of the case strikes against the ejector point and the case is ejected from the receiver.

d. Follower.—In magazine fire, after the last cartridge has been fired, the bolt will be locked in its open or rear position by the follower, which, under the action of the magazine spring, has been raised to block the bolt.

e. To unload.—To unload, move the bolt forward and back until no cartridges remain in the magazine or chamber.

f. Safety devices.—To set the rifle at "safe," turn the safety lock to the rear as far as it will go. This locks the bolt handle in position and lifts the sear notch off the sear nose.

g. Cocking.—The piece cannot be cocked except by actuation of the bolt. It is cocked by raising the bolt handle until it strikes the left side of the receiver, pulling the bolt backward until the sear notch engages the sear nose, then pushing it forward and turning it fully down. Raising the bolt handle half cocks the piece, and full cock is secured by pressure against the mainspring in forward movement of the bolt.

Section V

INDIVIDUAL SAFETY PRECAUTIONS

■ 22. Rules.—*a.* Consider every rifle to be loaded until you have examined it and proved it to be unloaded. Never trust your memory in this respect.

b. Never point the rifle at anyone you do not intend to shoot; never point it in a direction where an accidental discharge may do harm.

c. Always unload the rifle if it is to be left where someone else may handle it.

d. Always point the rifle up when snapping the trigger after examination.

e. If it is desired to carry the rifle cocked with a cartridge in the chamber, the bolt mechanism should be secured by turning the safety lock to the rear as far as it will go.

f. Under no circumstances should the firing pin be let down by hand or by manipulation of the trigger while closing the bolt on a cartridge in the chamber.

g. Never fire a rifle with any grease, dust, dirt, mud, snow, a cleaning patch, or other obstruction in the bore. To do so may burst the barrel.

h. Never fire a rifle with grease or oil on the ammunition or on the walls of the rifle chamber. This creates a hazardous pressure on the rifle bolt.

i. See that the ammunition is clean and dry. Examine all live and dummy ammunition. Turn in all cartridges which have loose bullets and those which appear to have other defects.

j. Do not allow the ammunition to be exposed to the direct rays of the sun for any length of time. Heat creates hazardous chamber pressure.

k. If the rifle misses fire, the bolt should not be opened or unlocked until enough time has elapsed to make sure that the rifle is not hanging fire. Since the rifle cannot be cocked except by opening the bolt, there is a temptation to open the bolt too soon. The bolt should not be opened for a full minute after a misfire. The precaution applies primarily to training; it is seldom practicable in combat conditions.

SECTION VI

SPARE PARTS AND ACCESSORIES

■ 23. SPARE PARTS.—*a*. The parts of any rifle will in time become unserviceable through breakage or wear resulting from continuous usage, and for this reason spare parts are supplied. These are extra parts provided for replacing the parts most likely to fail, for use in making minor repairs, and for the general care of the rifle. They should be kept clean and coated with a thin film of light preservative lubricating oil to prevent rust. Sets of spare parts should be kept complete at all times. Whenever a spare part is taken to replace a defective part in the rifle, the defective part should be repaired or a new one substituted in the spare parts set as soon as possible. Parts that are carried complete should at all times be correctly assembled and ready for immediate insertion in the rifle. The allowance of spare parts is prescribed for the rifle in SNL–B–4.

b. With the exception of replacements with the spare parts mentioned in *a* above, repairs or alterations to the rifle by using organizations are prohibited.

■ 24. APPENDAGES.—Appendages are items not required for use in the operation of the major equipment but are attached to or used in connection with this equipment. For the rifle they consist of the bayonet M1917 and bayonet scabbard M1917.

a. *Bayonet*.—The bayonet is a blade sharpened along the entire lower edge. The bayonet guard is constructed so as to fasten the bayonet securely to the rifle or its scabbard. Grips on both sides of the tang provide a handle when the bayonet is used as a hand weapon.

b. *Bayonet scabbard*.—The scabbard, shaped to receive the bayonet, consists of a leather body with a leather reinforcement at its tip which contains a drain hole. The scabbard is held to the belt of the soldier by two hooks.

■ 25. *Accessories*.—*a*. *General*.—Accessories include the tools required for assembling, disassembling, and cleaning the rifle; also the gun sling, spare parts containers, covers, arm locker, etc. Accessories should not be used for purposes other than those for which they are intended, and when not in use they should be stored in the places or receptacles provided for them. There are a number of accessories, the names or general characteristics of which indicate their uses or application. Accessories of a special nature or those which have special uses are described in *b* to *f*, inclusive, below.

b. *Arm locker and rack*.—The arm locker and the arm rack are used to store or stack rifles and pistols to prevent mishandling or pilfering.

c. *Brush and thong*.—The brush and thong are used for cleaning the bore of the rifle. The case in which they are carried is partitioned so that one end contains the oil and oil dropper and the other holds the tip, weight, thong, and brush.

d. *Cleaning rod M1 and cleaning brush M2*.—The cleaning rod M1 has a handle at one end and is threaded at the other end to receive the patch section or the brush. The cleaning brush M2 is used to clean the bore of the rifle.

e. Sling.—The sling, fastened to the loops of the rifle, may be adjusted to suit the individual. The sling consists of a long and a short strap, either of which may be lengthened or shortened as desired.

f. Ruptured cartridge extractor.—The ruptured cartridge extractor has the general form of a caliber .30 cartridge. It consists of three parts—the spindle, the head, and the sleeve. Before the ruptured cartridge extractor is used, the live cartridges must be removed from the rifle. The ruptured cartridge extractor is then inserted through the ruptured opening of the case and pushed forward into the chamber. The bolt is closed so that the extractor on the bolt engages the head of the ruptured cartridge extractor. The bolt is then drawn back; this action draws out the ruptured cartridge extractor, which holds the ruptured case on its sleeve.

SECTION VII

INSTRUMENTS

■ 26. BINOCULARS.—*a. General.*—There are two binoculars in use in the service: the M3 (fig. 10①) and the type EE (fig. 10②). The M3 will eventually replace the type EE. The subsequent paragraphs apply to both models. Each model consists of the binocular with its carrying case.

b. Description.—(1) The binocular consists of two compact prismatic telescopes (5) pivoted about a common hinge (4) which permits adjustment for interpupillary distances. A scale (3), graduated every 2 millimeters from 56 to 74, permits the observer to set the telescope rapidly to suit his eye distance when the spacing of his eyes is known. The eyepieces (1) can be focused independently for each eye by screwing them in or out. Each eyepiece is provided with a diopter scale (2) for rapid setting when the observer knows the correction for his eye. The zero graduations indicate the settings for normal eyes.

(2) The left telescope is fitted with a glass reticle (fig. 11) upon which are etched a vertical mil scale, a horizontal mil scale, and a stadia graduated similarly to the sight leaf graduation on the service rifle, but inverted.

c. Use.—The binocular is used for observations and the measurement of small horizontal and vertical angles in mils. The vertical stadia scale is used to pick up auxiliary aiming marks in direct laying and to determine troop safety for overhead fire.

d. Preliminary adjustments.—(1) *Interpupillary distances.*—To adjust the binocular so that the eyepieces are the same distance apart as the pupils of your eyes, point it

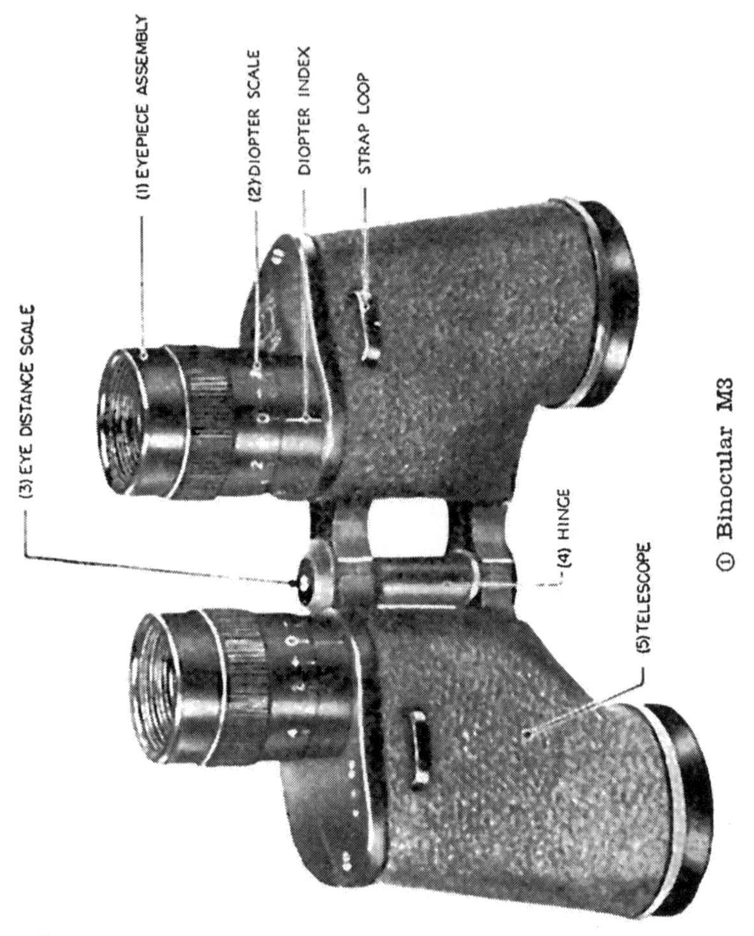

① Binocular M3

at the sky and open or close the hinged joint until the field of view ceases to be two overlapping circles and appears as one sharply defined circle. Note the reading on the scale

② Binocular, type EE.
FIGURE 10.—Binoculars.

(3), which indicates the spacing of your eyes. This setting on any other M3 or type EE binocular will accommodate your eyes.

(2) *Focus of eyepiece.*—Look through the binocular, with both eyes open, at an object several hundred yards away. Place the hand over the front of one telescope and screw the eyepiece of the other in or out until the object is sharply defined. Repeat this operation for the other eye and note the diopter scale. This setting on any other M3 binocular will accommodate your eyes.

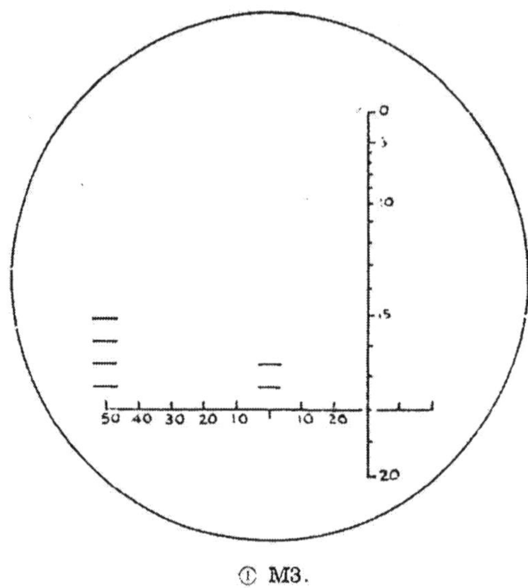

① M3.
FIGURE 11.—Reticles, binocular.

e. Operation.—(1) In using the binocular it should be held in both hands and pressed lightly to the eyes so as to keep the relation with the eyes constant without transmitting tremors from the body. The bent thumb should fit into the outer edges of the eye sockets in such a manner as to prevent light from entering in rear of the eyepieces. When possible it is best to use a rest for the binocular or for the elbows.

(2) The mil scales are seen when looking through the binocular and, by superimposing them upon the required objects, the horizontal and vertical angles between these objects may be read.

f. Care.—The binocular is a rugged, serviceable instrument but it should not be abused or roughly handled. Do not wipe the lenses with the fingers. Use only clean lens tissue paper

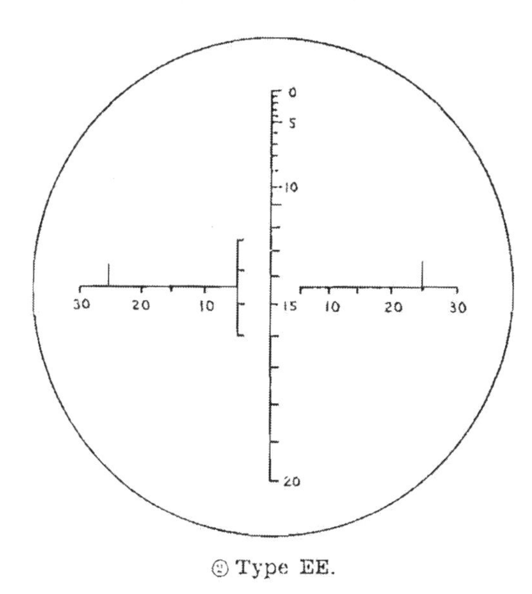

⑤ Type EE.

FIGURE 11.—Reticles, binoculars.—Continued.

to wipe the lenses if they become dirty. No substitute should ever be used for this purpose. The use of polishing liquids or pastes for polishing lenses is forbidden.

SECTION VIII

AMMUNITION

■ 27. GENERAL.—The information in this section pertaining to the several types of cartridges authorized for use in the U. S. rifle, caliber .30, M1917, includes description, means of identification, care, use, and ballistic data.

■ 28. CLASSIFICATION (figs. 12 and 13).—*a.* (1) Based upon use, the principal classifications of ammunition for this rifle are—

(*a*) *Ball*—for use against personnel and light matériel targets.

(*b*) *Tracer*—for observation of fire and incendiary purposes.

(*c*) *Armor-piercing*—for use against armored vehicles, concrete shelters, and similar bullet-resisting targets.

(2) Armor-piercing and tracer ammunition are similar to ball in outward appearance except that—

① Ball M2.

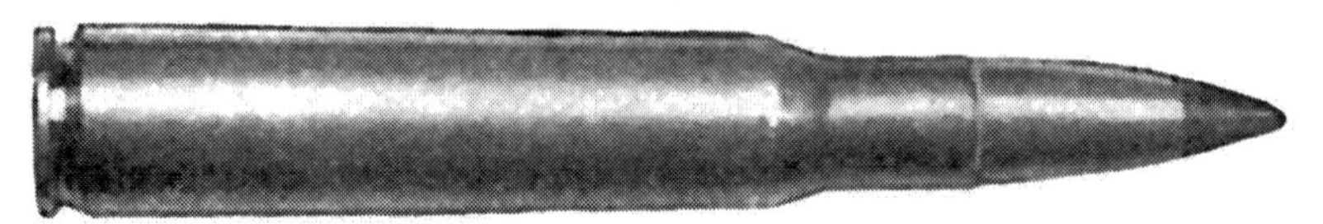

② Tracer M1.

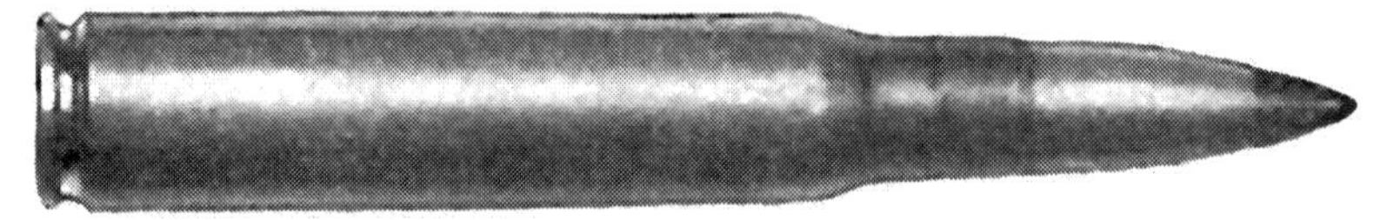

③ Armor piercing M1.

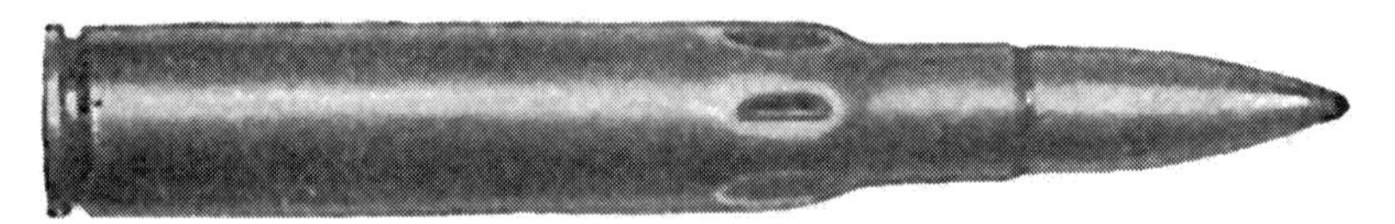

④ Guard M1906.

FIGURE 12.—Cartridges, caliber .30.

(*a*) Armor-piercing bullets are painted black for ¼ inch from the point.

(*b*) Tracer bullets are painted red for ¼ inch from point.

b. Other types provided for special purposes are—

(1) *Guard*—for guard purposes (gallery practice cartridges are also used for this purpose).

(2) *Blank*—for simulated fire, signaling, and salute.

(3) *Dummy*—for training. There are two types of dummy cartridges, practice and range (see fig. 13). The practice dummy cartridge is tinned and is made with six longitudinal corrugations; it has one circular hole in the corrugated portion. It may be used on the preparatory training field or on the range. The range dummy cartridge has the same appearance as ball cartridges, except that for purposes of identification it is provided with a longitudinal slot which starts

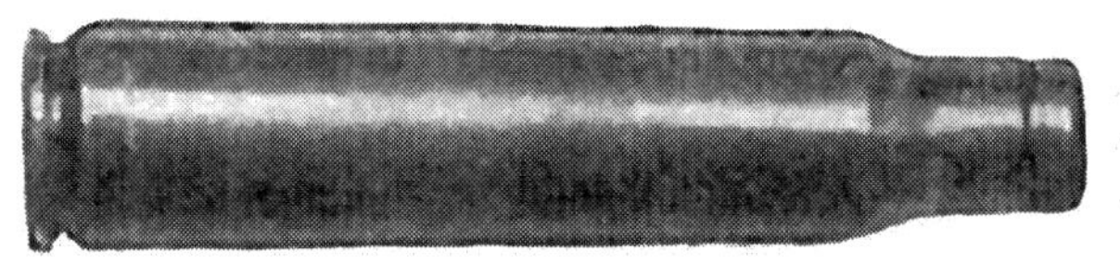

① Blank M1909.

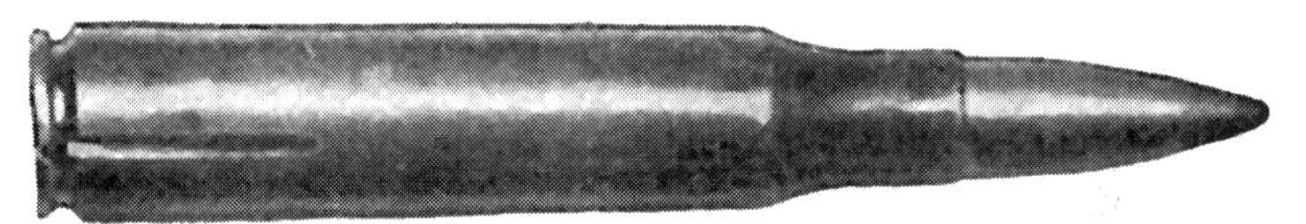

② Dummy range, slotted, M1921

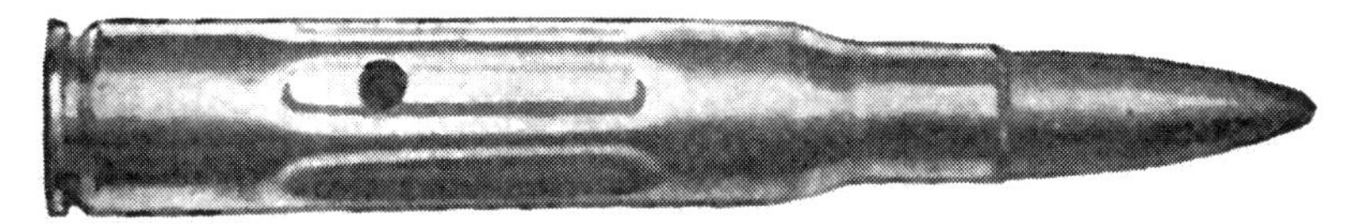

③ Dummy (corrugated) M1906.

FIGURE 13.—Cartridges, blank and dummy, caliber .30.

from the groove and extends about ¾ inch along the case. *The use of the range dummy is limited to the firing line on the range. Its use for practice on other practice fields or areas is strictly prohibited.*

(4) *Grenade, AT*—for use against tanks and other armored targets.

(5) *Grenade, AT, practice*—for training purposes.

■ 29. LOT NUMBER.—When ammunition is manufactured an ammunition lot number is assigned. This lot number becomes an essential part of the marking in accordance with specifications, and is marked on all packing containers and on the identification card inclosed in each packing box. It is required for all purposes of record, including grading and use reports on condition, functioning, and accidents in which the ammunition might be involved. Only those lots which are assigned grades appropriate for the weapon will be fired. Since it is impracticable to mark the ammunition lot number on each cartridge, every effort should be made to retain the ammunition lot number with the cartridges once the cartridges are removed from their original packing. Cartridges which have been removed from the original packing and for which the ammunition lot number has been lost are placed in grade 3. Therefore, it is obvious that when cartridges are removed from their original packings, they should be so marked that the ammunition lot number is preserved.

■ 30. GRADE AND IDENTIFICATION.—Current grades of all existing lots of small-arms ammunition are established by the Chief of Ordnance and are published in Ordnance Field Service Bulletin No. 3-5. No lot other than that appropriate to the weapon will be fired. Colored bands painted on the sides and ends of the packing boxes further identify the various types of ammunition. The following colored bands are used:

Armor-piercing	Blue on yellow.
Ball	Red.
Blank	Blue.
Dummy	Green.
Gallery practice	Brown.
Guard	Orange.
Tracer	Green on yellow.
Rifle grenade	Two blue.

■ 31. CARE, HANDLING, AND PRESERVATION.—*a*. Small-arms ammunition is not dangerous to handle. However, care must be observed to keep the boxes from becoming broken or damaged. All broken boxes must be repaired immediately and care taken that all markings are transferred to the new parts of the box. The metal liner should be air tested and sealed if equipment for this work is available.

b. Ammunition boxes should not be opened until the ammunition is required for use. Ammunition that is removed from the airtight container, particularly in damp climates, is likely to corrode, and thus become unserviceable.

c. Ammunition should be protected from mud, sand, dirt, and water. Ammunition that gets wet or dirty should be wiped off at once. Light corrosion on cartridges should be wiped off. However, cartridges should not be polished to make them look better or brighter.

d. No caliber .30 ammunition, other than blanks, will be fired until it has been positively identified by ammunition lot number and grade.

■ 32. CARTRIDGE, BALL, CALIBER .30, M2 AND M1.—The maximum range of the M2 cartridge is about 3,450 yards; for the M1 cartridge about 5,500 yards. The M2 type is standard.

■ 33. PRECAUTIONS IN FIRING BLANK AMMUNITION.—*a*. It is dangerous to fire blank cartridges at personnel at distances less than 20 yards. The wad or paper cup may fail to break up.

b. Misfires in which the primer explodes but fails to ignite the powder charge may prove dangerous when blank ammunition is being fired. In this type of misfire, some of the powder may be blown into the bore of the weapon. A series of such rounds in which the powder fails to ignite will result in an accumulation of powder sufficient to cause serious damage when ignited by a normal cartridge. When misfires in excess of 5 percent are encountered in blank ammunition, the firing of the lot must be suspended and report made to the Chief of Ordnance.

CHAPTER 2

MARKSMANSHIP—KNOWN-DISTANCE TARGETS

SECTION I

GENERAL

■ 34. PURPOSE.—The purpose of this chapter is to provide a thorough and uniform method of training individuals to be good rifle shots and of testing their proficiency in firing at known-distance targets.

■ 35. NECESSITY FOR TRAINING.—*a.* Without proper training a man instinctively does the wrong thing in firing the rifle. He gives the trigger a sudden pressure which causes flinching. However, if he is thoroughly instructed and drilled in the mechanism of correct shooting, and is carefully and properly coached when he begins firing, he rapidly acquires correct shooting habits.

b. Rifle firing is a mechanical operation. Anyone who is physically and mentally fit to be a soldier can learn to do it well if properly instructed and trained. The methods of instruction are the same as those used in teaching any mechanical operation. The training is divided into steps which must be taught in proper sequence. The soldier is carefully coached and is corrected whenever he starts to make a mistake.

■ 36. FUNDAMENTALS.—To become a good rifle shot the soldier must be thoroughly trained in the following essentials of good shooting:

a. Correct sighting and aiming.

b. Correct positions.

c. Correct trigger squeeze.

d. Correct application of rapid fire principles.

e. Knowledge of proper sight adjustments.

■ 37. PHASES OF TRAINING.—*a.* Marksmanship training is divided into two phases: preparatory training and range practice.

b. No soldier should be given range practice until he has had a thorough course in preparatory training.

c. The soldier should be proficient in mechanical training and related subjects before he receives instruction in marksmanship training.

d. Every man who is to fire on the range is put through the entire preparatory course. No distinction is made between recruits and men who have had range practice, regardless of their previous qualification. Some part of the preparatory instruction may have escaped the men in previous years; it is certain that some of it has been forgotten. In any event it is helpful to go over it again and refresh the mind on the subject.

e. All of the noncommissioned officers and other men selected as assistant instructors and special coaches of the unit are put through a course of instruction and required to pass a rigid test before the marksmanship training for the unit begins.

■ 38. PRACTICE SEASONS.—*a. Regular.*—(1) The provisions of this paragraph apply to peacetime conditions. Ordinarily the regular practice season for the Regular Army should cover a period of not more than 3 weeks for each unit. A period of 1 week should be devoted to preparatory exercises and 2 weeks to range practice. When unforeseen circumstances interrupt the period of instruction, the time may be extended by the post commander.

(2) The regular practice season for units of the National Guard and other components of the Army will be of such duration and ordered at such times as may be best suited for effective training.

(3) Battalions or smaller units should be relieved from routine garrison duty during the period of preparatory training and range practice.

(4) In order that officers and enlisted men of the Regular Army, National Guard, and Organized Reserves may be familiar with their exact duties in marksmanship training during mobilization, units should not exceed the number of hours

set forth in the model schedule given in paragraph 158. Under no conditions will any man be given range practice until he has had a thorough course in the preparatory exercises.

b. Supplementary.—Supplementary practice is not necessary if the regular practice season has been efficiently conducted except when a large number of unqualified men join the unit after the regular practice season. The supplementary practice season is usually placed as late in the fall as is consistent with efficient instruction. However, this practice may be held at any time when circumstances make it advisable.

■ 39. CONTINUOUS PRACTICE.—Rifle practice is not limited to a particular season. Subject to ammunition allowances, commanding officers will adopt such measures as may be necessary to maintain a high state of excellence in rifle firing throughout the year. The particular measures adopted will depend upon the facilities near the post or station. The measures taken may provide for competitions between individuals or units, or the encouragement of small-bore rifle teams. Regularly scheduled periods of supervised dry shooting are an excellent means of maintaining a high state of training.

■ 40. RECRUIT INSTRUCTION.—As part of their recruit training, all recruits armed with the rifle will be thoroughly instructed in mechanical training and in the fundamental elements of rifle marksmanship—sighting and aiming, positions, trigger squeeze, and rapid fire. Instruction in rifle marksmanship will begin with the initial instruction of the recruit and will continue throughout the period of recruit training.

■ 41. LEADERS AND COMMANDERS.—*a. Duties.*—(1) *Squad leader.*—(*a*) Organizes the work in his squad so that each man is occupied during the preparatory period in the prescribed form of training for target practice.

(*b*) Tests each man in his squad at the end of the training on each preparatory step and assigns him a mark in the proper place on the blank form showing state of training.

(*c*) Sees that each man takes proper care of his rifle and cleans it at the end of each day's firing.

(*d*) Requires correct aiming, correct positions, and proper trigger squeeze when fire is simulated in drills and maneuvers.

(2) *Platoon leader.*—Supervises and directs the squad leaders in training their squads; personally checks each man in his platoon on the points enumerated on the blank form; and examines each man as indicated in paragraph 49.

(3) *Company commander.*—Requires the prescribed methods of instruction and coaching to be carried out in detail; supervises and directs the squad and platoon leaders; in small companies that lack qualified platoon leaders, he performs the duties prescribed for platoon leaders in (2) above.

(4) *Battalion commander.*—Sees that his instructors know the prescribed methods of instruction and coaching; supervises the instruction within his battalion and requires his instructors to follow the preparatory exercises and methods of coaching in detail.

b. Equipment.—All equipment used in the preparatory exercises must be accurately and carefully made. One of the objects of these exercises is to cultivate a sense of exactness and carefulness in the minds of the men undergoing instruction. They cannot be exact with inexact instruments, and they will not be careful when working with equipment that is carelessly made.

SECTION II

PREPARATORY MARKSMANSHIP TRAINING

■ 42. GENERAL.—*a.* The purpose of preparatory marksmanship training is to teach the soldier the essentials of good shooting, and to develop fixed and correct shooting habits before he undertakes range practice.

b. Preparatory marksmanship training is divided into six steps:

(1) Sighting and aiming exercises.

(2) Position exercises.

(3) Trigger-squeeze exercises.

(4) Rapid-fire exercises.

(5) Instruction in the effect of wind, in sight changes, and in the use of the score book.

(6) Examination of men before starting range practice.

c. Each step is divided into exercises designed to teach the soldier the importance of each operation and to drill him in those operations until he is able to execute them correctly.

d. Instruction in the effect of wind, sight changes, and the use of the score book can be taught indoors during inclement weather. It is not a training step that need be given in any particular sequence but it must be covered prior to the examination. The first four steps are taught in the order listed; each one involves the technique learned in the preceding steps.

e. Each of the first four steps starts with a lecture by the instructor to the assembled group. This lecture includes a demonstration in which the instructor directs a squad in the execution of each exercise. He shows exactly how to do the exercises that are to be taken up, explains why they are done and their application to rifle shooting. He shows how the squad leader organizes the work so that no men are idle, and how the men coach each other when they are not under instruction by an officer or noncommissioned officer. These talks and demonstrations are an essential part of the training. If properly given, they awaken the interest and enthusiasm of the whole command for the work and give an exact knowledge of how each step is to be carried out—something that men cannot get from reading a description, no matter how accurate and detailed that description may be. These talks and demonstrations may be given by the platoon leader, the company commander, or the battalion commander to their respective units; or the instruction may be given by a specially qualified officer who has been detailed as officer in charge of rifle instruction. The instructors who apply the demonstrated exercises to the men of the command are the officers and noncommissioned officers of the units undergoing instruction.

f. The following form showing the state of instruction should be kept by each squad leader and by each platoon leader independently of his squad leaders:

Name	Care and cleaning of rifle	Sighting bar	Sighting and aiming with rifle	Shot group exercises	Blackening sights	Use of gun sling	Taking up slack	Holding breath	Positions	Assuming positions rapidly	Trigger squeeze	Calling shot	Bolt operation	Rapid fire	Sight changes	Effect of light and wind	Use of score book	Ability to coach	Final examination	Gallery practice	Pads

METHOD OF MARKING

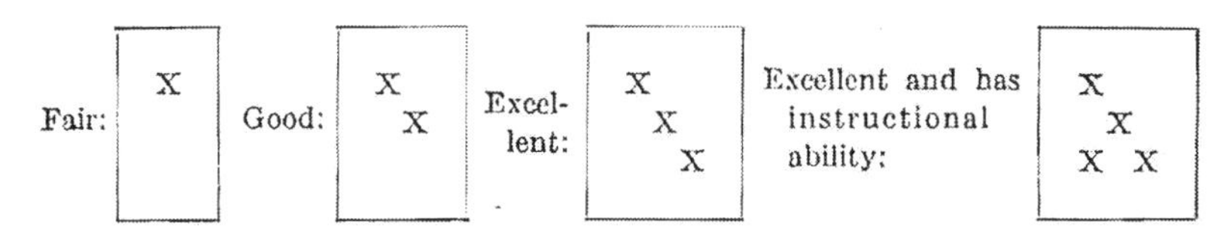

g. The instruction must be thorough and it must be individual. Each man must understand every point and be able to explain it in his own words. The company commander carefully supervises the work. From time to time he should pick men at random from the different platoons and put them through a test to see if the instruction is thorough and is progressing satisfactorily.

h. Interest and enthusiasm must be sustained. If these exercises become perfunctory, they do more harm than good.

i. Each man is tested thoroughly before he is allowed to fire (see par. 49).

j. During the preparatory exercises, whenever a man is in a firing position, the coach-and-pupil system is used.

The men are grouped in pairs and take turns in coaching each other. The man undergoing instruction is called the pupil. The man giving instruction is called the coach. When the men of a pair change places, the pupil becomes the coach and the coach becomes the pupil.

k. Correct shooting habits should be acquired during the preparatory training period. All errors must be noted, brought to the attention of the pupil, and corrected. Each soldier must be impressed with the importance of exactness in every detail. For example, there is no such thing as a trigger squeeze that is *about right;* it is either perfect or it is wrong.

l. Only practice dummy ammunition will be used during the preparatory training. The use of range dummy ammunition is prohibited except on the firing line.

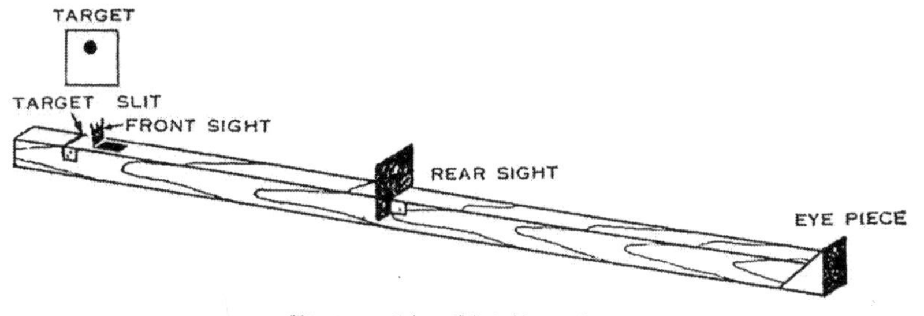

FIGURE 14.—Sighting bar.

■ 43. BLACKENING SIGHTS.—In all preparatory exercises involving aiming, and in all range firing, both sights of the rifle should be blackened. Before the sights are blackened, they should be cleaned and all traces of oil removed. The blackening is done by holding each sight for a few seconds in the point of a small flame so that a uniform coating of lampblack will be deposited on the metal. Materials commonly used for this purpose are carbide lamp, cylinder of carbide gas, kerosene lamp, candles, small pine sticks. Shoe paste may also be used. Carbide gas from a cylinder or a lamp is the most satisfactory of the materials named.

■ 44. FIRST STEP: SIGHTING AND AIMING.—*a. First exercise.*— The instructor displays a sighting bar (fig. 14) before his group and explains its use as follows:

(1) The front and rear sights on the sighting bar represent enlarged rifle sights.

(2) The sighting bar is used in the first sighting and aiming exercise because with it small errors can be easily seen and explained to the pupil.

(3) The eyepiece requires the pupil to place his eye in such a position that he sees the sights in exactly the same alinement as the coach sees them.

(4) There is no eyepiece on the rifle, but the pupil learns by use of the sighting bar how to aline the sights when using the rifle.

(5) The removable target attached to the end of the sighting bar is a simple method of readily alining the sights on a bull's-eye.

(6) The instructor explains the peep sight to the assembled group and shows each man the illustrations of a correct sight alinement (fig. 15).

(7) With the target removed, the instructor adjusts the sights of the sighting bar to illustrate a correct alinement of the sights. Each man of the assembled group looks through the eyepiece at the sight adjustments.

(8) He adjusts the sights of the sighting bar with various small errors in sight alinement and has each man of the assembled group try to detect the error.

(9) The instructor describes a correct aim, showing the illustration to each man. He explains that the top of the front sight is seen through the middle of the circle and just touches the bottom of the bull's-eye, so that all the bull's-eye can be clearly seen (fig. 15).

(10) The eye should be focused on the bull's-eye in aiming, and the instructor assures himself, by questioning the men, that each understands what is meant by focusing the eye on the bull's-eye.

(11) The instructor adjusts the sights of the sighting bar and the removable target so as to illustrate a correct aim and requires each man of the group to look through the eyepiece to observe the correct aim.

(12) He adjusts the sights and the removable target of the sighting bar so as to illustrate various small errors and requires each man in the group to attempt to detect the error.

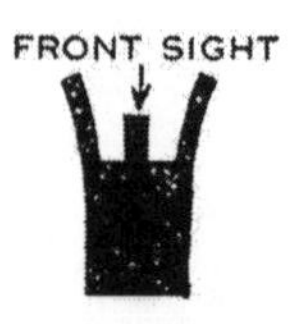

FIGURE 15.—Sight alinement and aim.

(13) The exercise described above having been completed by the squad leader or other instructor, the men are placed in pairs and repeat the exercise by the coach-and-pupil method.

(14) As soon as the pupil is considered proficient in the first sighting and aiming exercise, he is put through the second and third sighting and aiming exercises by the instructor. Such pupils are then placed in pairs to instruct each other in these two exercises by the coach-and-pupil method.

b. Second exercise.—(1) A rifle with sights blackened is placed in a rifle rest and pointed at a blank sheet of paper mounted on a box (fig. 16). Without touching the rifle or rifle rest, the coach takes the position illustrated and looks through the sights. The coach directs the marker by command or improvised signal to move the small disk until the bottom of the bull's-eye is in correct alinement with the sights and then commands: HOLD, to the marker. The coach moves away from the rifle and directs the pupil to look through the sights to observe the correct aim.

(2) After the pupil has observed the correct aim, the marker moves the disk out of alinement. The pupil then takes position and directs the marker to move the disk until the bottom of the bull's-eye is in correct alinement with the sights. The coach then looks through the sights to see if the alinement is correct.

FIGURE 16.—Position for second sighting and aiming exercise.

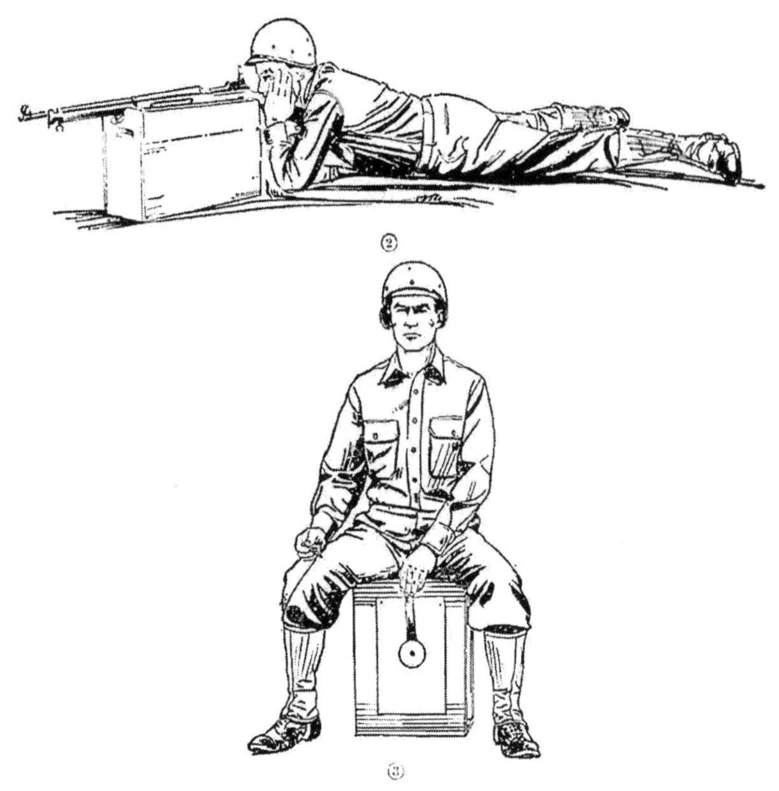

FIGURE 16.—Position for second sighting and aiming exercise—Continued.

(3) The coach alines the sights on the bull's-eye with various slight errors to determine whether the pupil can detect them.

c. Third exercise.—(1) The object of this exercise is to show the importance of uniform and correct aiming, and to foster a sense of exactness.

(2) This exercise is conducted as follows: The rifle with the sights blackened is placed in a rifle rest and pointed at a blank sheet of paper mounted on a box. The pupil takes the position illustrated and looks through the sights without touching the rifle or rifle rest. The pupil directs the marker to move the disk until the bottom of the bull's-eye is in correct alinement with the sights, and then command: HOLD, to the marker. The coach then looks through the sights to see if the alinement is correct. Without saying anything

to the pupil, he commands: MARK, to the marker. The marker, without moving the disk, makes a dot on the paper with a sharp-pointed pencil inserted through the hole in the center of the bull's-eye. The marker then moves the disk to change the alinement. The pupil and coach without touching the rifle or rifle rest repeat this operation until three dots, numbered 1, 2, 3, respectively, have been made. These dots outline the shot group, and the pupil's name is written under it. The size and shape of the shot group are then discussed and the errors pointed out. This exercise is repeated until proficiency is attained. At 50 feet and with a small bull's eye a man should be able to place all three marks so that they can be covered by the unsharpened end of a lead pencil.

(3) This exercise should also be practiced at 200 yards on a 10-inch movable bull's eye (fig. 17), and if time permits at 500 yards on a 20-inch movable bull's eye. These long-range shot group exercises teach the men to aim accurately at a distant target the outlines of which are indistinct. If the exercise is properly handled, it helps greatly to sustain interest in the work. At 200 yards a man should be able to make a shot group that can be covered with a silver dollar, and at 500 yards a shot group not over 2 inches in diameter.

(4) Tissue paper may be used to trace each man's shot group at the long ranges. The name of the pupil is written on the tissue paper under the shot group he made. These tracings are sent back to the firing line so that the pupil can see what he has done.

(5) The third sighting and aiming exercise, especially the 200 yard shot group work, is continued during the time devoted to the second and third preparatory steps. The reason for continuing this exercise is to bring backward men up to the required state of proficiency and to maintain interest.

(6) Competition between the individuals of a squad to see who can make the smallest shot group creates interest.

■ 45. SECOND STEP: POSITIONS.—*a. General.*—Instruction in positions includes the use of the gun sling, taking up the slack in the trigger, holding the breath while aiming, aiming, and the use of the aiming device.

FIGURE 17.—Position for third sighting and aiming exercise on long ranges.

b. Gun sling.—(1) The gun sling, properly adjusted, is of great assistance in shooting in that it helps to steady the rifle. Each man is assisted by the instructor in securing the correct adjustment for his sling. In a firing position the sling should be adjusted to give firm support without discomfort to the soldier. The gun sling is readjusted for drill purposes by means of the lower loop without changing the adjustment of the upper loop.

(2) There are two authorized adjustments—the loop sling and the hasty sling. The hasty sling is more rapidly adjusted than the loop sling, but it gives less support in positions other than the standing position.

(*a*) *Loop adjustment* (fig. 18).

1. Loosen the lower loop.

2. Insert the left arm through upper loop from right to left, so that the upper loop is near the shoulder and well above the biceps muscle.

3. Pull the keepers and hook close against the arm to keep the upper loop in place.

①

FIGURE 18.—Loop sling adjustment.

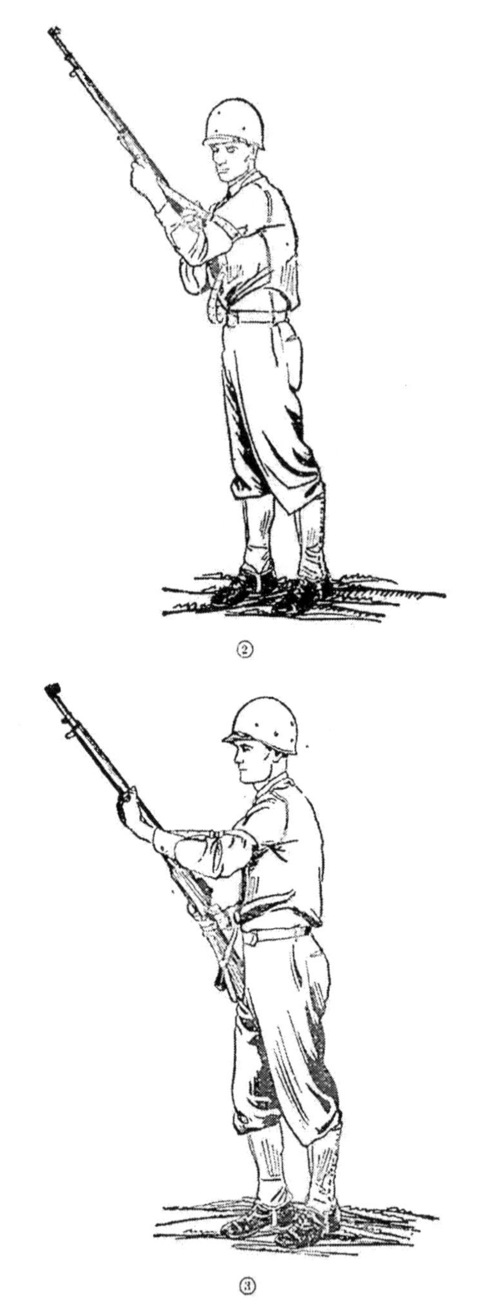

②

③

FIGURE 18.—Loop sling adjustment—Continued.

4. Move the left hand over the top of the sling and grasp the rifle near the lower band swivel so as to cause the sling to lie smoothly along the hand and wrist. The lower loop, not used in this adjustment, should be so loose as to prevent any pull upon it. Neither end will be removed from either swivel.

(*b*) *Hasty sling adjustment* (fig. 19.)

1. Loosen the lower loop.

2. Grasp the rifle just behind the lower band swivel with the left hand and grasp the small of the stock with the right hand.

3. Throw the sling to the left and catch it above the elbow and high on the arm.

4. Remove the left hand from the rifle, pass the left hand under the sling, then over the sling and regrasp the rifle with the left hand so as to cause the sling to lie along the hand and wrist. The sling may be given one-half turn to the

①

FIGURE 19.—Hasty sling adjustment.

FIGURE 19.—Hasty sling adjustment—Continued.

left and then adjusted. This twisting causes the sling to lie smoothly along the hand and wrist.

c. Taking up slack.—The first movement of the trigger which takes place when light pressure is applied is called taking up the slack. It is part of the position exercise because this play must be taken up by the finger as soon as the correct position is assumed and before careful aiming is begun. The entire amount of slack in the trigger is taken up by one positive movement of the finger.

d. Holding breath.—(1) Holding the breath in the proper manner while aiming is important. It will be found that a large proportion of men in any group undergoing instruction in rifle practice do not know how to hold the breath in the proper manner. Each man must be carefully instructed and tested on this point. The correct manner of holding the breath must be practiced at all times during position and trigger-squeeze exercises and whenever firing or simulating fire.

(2) To hold the breath properly draw into the lungs a little more air than is used in an ordinary breath. Let out a little of this air and stop the remainder by closing the throat so that the air remaining in the lungs will press against the closed throat. Do not hold the breath with the throat open or by the muscular action of the diaphragm as if attempting to draw in more air. The important point is to be comfortable and steady while aiming and squeezing the trigger.

e. Aiming.—The rifle is carefully aimed at a target each time a firing position is assumed. The aiming device may be used by the coach to check the aim.

f. General rules for positions.—The general rules which follow are common to the prone, sitting, kneeling, and standing positions. The exact details of a position depend upon the conformation of the individual.

(1) To assume any position, half face to right and then assume the position.

(2) In assuming any position there is some point to which the rifle points naturally and without effort. If this point is not the center of the target the whole body must be shifted

so as to bring the target into proper alinement with the sights. Otherwise the firer will be firing under a strain because he will be pulling the rifle toward the target by muscular effort for each shot.

(3) The right hand grasps the small of the stock. The right thumb may be either around the small of the stock or along the right side of the stock.

(4) The left hand is against or near the lower band swivel, the hand and wrist joint straight, rifle resting in the crotch formed by the thumb and index finger and resting on the base of the thumb and heel of the palm of the hand.

(5) The left elbow is placed as nearly under the rifle as it can be without appreciable effort.

(6) Ordinarily the second joint of the index finger contacts the trigger. The first joint may be used by men the length of whose arm or the size of whose hand is such as to make it difficult to reach the trigger with the second joint, or to whom the first joint of the finger seems more natural and comfortable.

(7) The cheek is pressed firmly against the stock and placed as far forward as possible without strain to bring the eye near the rear sight.

(8) The butt of the rifle is held firmly against the shoulder.

(9) The rifle should not be canted.

(10) Left-handed men who have difficulty with the right-hand position will be allowed to use the left-hand position.

(11) The rifle is held by the dead weight of the body and tension of the sling. No muscular effort should be exerted. Muscular relaxation is important in steady holding. Only the muscles that operate the trigger finger are involved in firing the shot.

(12) The positions are the same when no sling is used, except that some muscular effort of the left arm and hand is necessary, and the right hand grips the small of the stock more firmly.

g. Prone position (fig. 20).—(1) In this position, the body lies at an angle of about 45° to the line of aim with the spine straight; legs are well apart; inside of the feet should be flat on the ground, or as nearly so as is possible without strain; elbows are well under the body so as to raise the chest off the ground; right hand grasps the small of the stock; left hand is near the lower band swivel, wrist straight, rifle placed in the crotch formed by the thumb and index finger and resting on the heel of the hand; cheek pressed firmly against the stock with the eye as near the cocking piece as possible without strain; gun sling properly adjusted and tight enough to give firm support, butt of the rifle pressed firmly against the shoulder.

FIGURE 20.—Prone position.

(2) The elbows should not be unduly spread apart because this results in an unsteady position and brings the chest so near the ground that the neck has to be strained backward in order to see through the sights. This strained position of the neck interferes with good vision and tends to make the firer unsteady. The exact angle of the arms to the ground will depend upon the conformation of the man. The right elbow should be so placed that the right upper

arm will not form an angle of less than 45° with the ground.

(3) The exact position of the left hand depends on the length of arm and width of chest of the individual. It should be as near the lower band swivel as the conformation of the man permits.

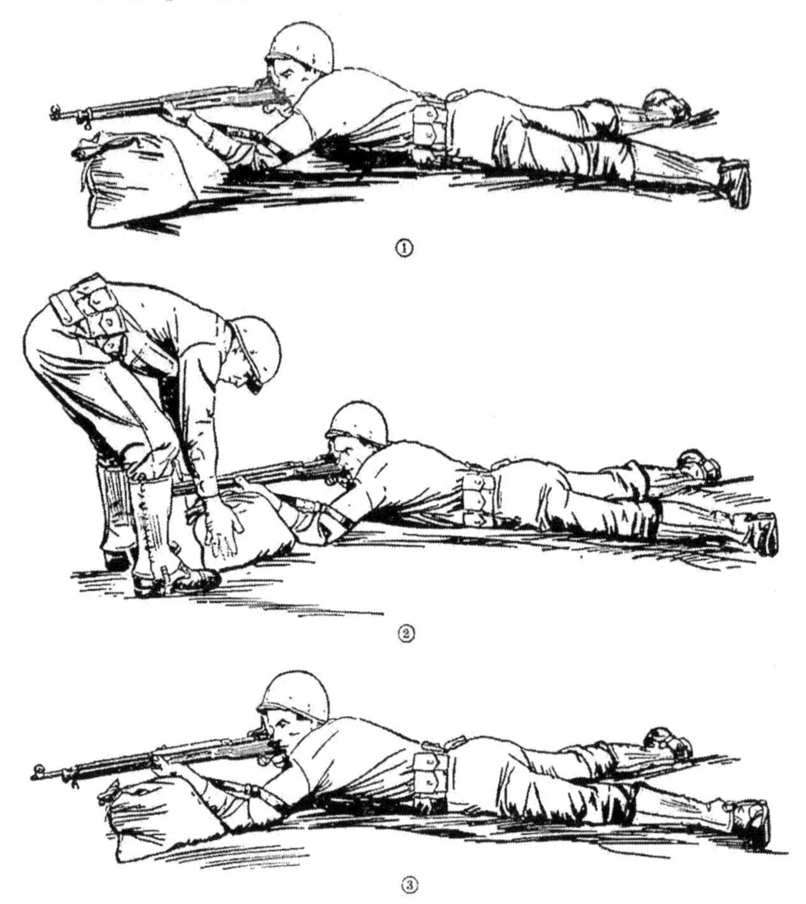

FIGURE 21.—Prone position with sandbag rest.

h. Sandbag rest position (fig. 21).—(1) The sandbag rest position conforms in every detail to the normal prone position described above, with the addition of a sandbag which supports the left forearm, wrist, and hand.

(2) The bag is a little more than half full and tied near the top so as to leave considerable free space above the sand.

(3) It is important that the sandbag be high enough to permit the taking of the normal prone position. The natural tendency is to have a low rest and to be very flat on the ground with the elbows spread apart. This faulty position will cause scores to be lower than they would be if no rest at all were used. The properly adjusted sandbag is a great help. When it is not properly adjusted it is a handicap.

(4) The sandbag rest position is used in the first stages of a pupil's training, not to teach steadiness of holding but to teach the correct trigger squeeze. By using the sandbag the slight unsteadiness of the hold is avoided, and the temptation to try to snap in the shot at the instant the sights touch the bull's-eye is eliminated.

(5) The coach will adjust the sandbag as follows:

(*a*) Have the pupil assume the prone position and aim at the target.

(*b*) Set the sandbag on its bottom and arrange the sand so that it is slightly higher than the back of the pupil's left hand.

(*c*) Facing the pupil, straddle the rifle barrel, and slide the sandbag against the pupil's left forearm, so that the narrow side of the bag supports his forearm and wrist and the back of his hand rests on top.

(*d*) Lower the sandbag to the proper height by pounding it with the hand.

i. Sitting position (fig. 22).—(1) The firer sits half-faced to the right; feet well apart and well braced on the heels, which are dug slightly into the ground; body leaning well forward from the hips with back straight; both arms resting inside the legs and well supported; cheek pressed firmly against the stock and placed as far forward as possible without straining; left hand near the lower band swivel, wrist straight, rifle placed in the crotch formed by the thumb and index finger and resting on the heel of the hand.

(2) The sitting position is used in the field when firing from ground that slopes downward to the front. In practicing this position the feet may be slightly lower than the ground upon which the pupil sits. Sitting on a low sandbag is authorized.

FIGURE 22.—Sitting position.

①

FIGURE 23.—Kneeling position.

(3) In the event the conformation of a man is such that he cannot assume the prescribed normal position, such changes as may be necessary to secure a steady, comfortable position are authorized.

j. Kneeling position (fig. 23).—The firer kneels half-faced to the right on the right knee, sitting on the right heel; the left knee bent so that the left lower leg is vertical (as seen from the front); left arm well under the rifle and resting

②

FIGURE 23.—Kneeling position—Continued.

on the left knee with the point of the elbow beyond the knee-cap; right elbow above or at the height of the shoulder; cheek pressed firmly against the stock and placed as far forward as possible without strain. Sitting on the side of the foot instead of the heel is authorized.

k. Standing position (fig. 24.).—The firer stands half-faced to the right; feet from 1 foot to 2 feet apart; body erect and well balanced; left elbow well under the rifle; left hand in front of the balance, wrist straight, rifle placed in the crotch formed by the thumb and index finger and resting on the heel of the hand; butt of the piece high up on the shoulder and firmly held; right elbow approximately at the height of the shoulder; cheek pressed against the stock and placed as far forward as possible without strain. A position with the left hand against or under the trigger guard and with the

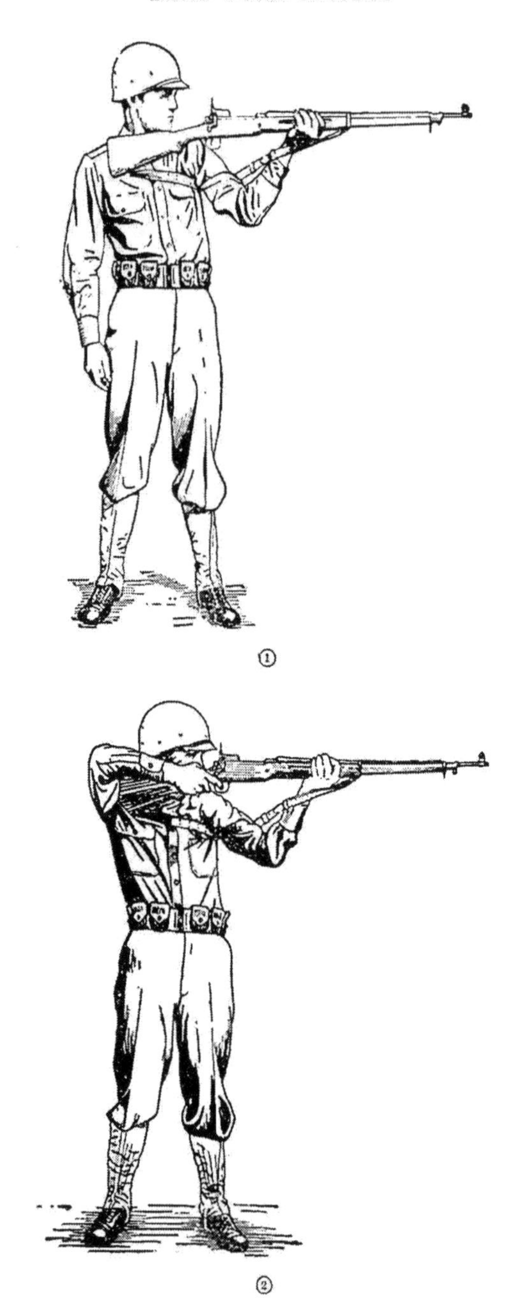

FIGURE 24.—Standing position.

left upper arm supported against the body is not a practical field position and is prohibited.

l. Procedure in conducting position exercises.—(1) Aiming points should be provided for position exercises. Small bull's-eyes may be used. They should be placed at a range of 1,000 inches and at different heights so that in aiming from various positions the rifle will be nearly horizontal. Standard known-distance targets may also be used as aiming points. They should be installed at distances used on the known-distance range.

(2) Before taking up each phase of the position exercises the instructor assembles his squad or group, and—

(*a*) Shows them the proper method of blackening the front and rear sights of the rifle and has each pupil blacken his sights.

(*b*) Explains and demonstrates the hasty sling adjustment and helps each pupil adjust his sling. He explains the loop sling adjustment and helps each pupil adjust his sling.

(*c*) Explains and demonstrates the proper manner of taking up the slack and has each pupil practice it.

(*d*) Explains and demonstrates the proper manner of holding the breath and has each pupil practice it.

(*e*) Explains and demonstrates the use of the aiming device.

(*f*) Explains the general rules which apply to all positions.

(*g*) Explains and demonstrates the different positions.

(3) After these explanations and demonstrations the instruction becomes individual by the coach-and-pupil method. Each pupil, after seeing that his sights are blackened, adjusts his sling, takes position, alines his sights, takes up the slack, holds his breath, and perfects the sight picture. As soon as his aim becomes unsteady, the exercise ceases. After a short rest the pupil repeats the exercise without further command. The trigger is not squeezed in the position exercise. Exercises are conducted in all positions.

(4) In the position exercises the coach sees that—

(*a*) The sights are blackened.

(*b*) The gun sling is properly adjusted and is tight enough to give support.

(*c*) The proper position is taken.

(*d*) The pupil is fully relaxed, holding the rifle by dead weight and not by muscular effort.

(*e*) The head and right hand are firmly fitted to the stock.

(*f*) The breath is properly held.

(*g*) The slack is taken up promptly.

(*h*) The pupil aims.

(*i*) The breath is held while aiming.

(*j*) The pupil does not hold and aim so long as to become uncomfortable and unsteady.

(*k*) The coach checks the pupil's manner of holding his breath by watching his back. The pupil's aim may be checked occasionally by means of the aiming device.

■ 46. THIRD STEP: TRIGGER SQUEEZE.—*a. Importance of trigger squeeze.*—(1) The most important item in rifle shooting is to squeeze the trigger in such a way as to fire the rifle without affecting the aim. Misses and poor shots are caused by spoiling the aim just before the discharge takes place. This is the result of jerking the trigger and flinching. The trigger must be squeezed so steadily that the firer cannot know the instant the piece will be fired. If a man squeezes the trigger so steadily that he cannot know when the discharge will take place, he does not spoil his aim and he will not flinch, because he does not know when to flinch.

(2) No good shot attempts to discharge the piece instantly upon alining his sights on the mark. He holds his aim as accurately alined on the mark as possible and maintains a steadily increasing pressure upon the trigger until the shot is fired. This method of squeezing the trigger must be carried out in all simulated firing or the value of the practice is lost.

(3) There is only one correct method of squeezing the trigger—a steady increase of pressure so that the firer does not know when the discharge will take place.

(4) Expert shots are men who through training have learned to increase the pressure only when the sights are in correct alinement with the bull's-eye. When the sights move slightly out of alinement, they hold what they have with the finger and only continue the increase of pressure when the sights are again properly alined.

(5) The difference between poor shots and good shots is measured in their ability to squeeze the trigger properly. Any

man with fair eyesight and strength can aline the sights on the target and hold them there for an appreciable length of time. When he has acquired sufficient will power and self-control to forget that there is to be an explosion and a shock, and squeezes the trigger with a steady increase of pressure until the rifle is fired, he has become a good shot, and not until then. This squeeze of the trigger applies to rapid fire as well as slow fire. The increase of pressure is faster in rapid fire but the process is the same.

b. Calling shot.—The pupil must always notice where the sights are pointed at the instant the rifle is fired, and must call out at once where he thinks the bullet will hit. Shots are called even when simulating fire at a mark, in order to acquire the habit and to develop a closer hold. No man is a good shot until he can call his shot before it is marked. Inability to call a shot indicates the firer did not know where the sights were pointing at the time the rifle was fired; in other words, he shut his eyes first and fired afterward.

c. Trigger-squeeze exercise.—(1) *General.*—(*a*) The instructor explains to the assembled squad or group the importance of correct trigger squeeze. He assures himself by questions that each pupil understands what is meant by a steady increase of pressure; that is, that the increase is applied with a steady, smooth pressure and not by a sudden jerk or series of impulses.

(*b*) The pupil is first taught the trigger squeeze in the prone position with the sandbag rest. In this position he can hold steadily and has not the temptation to *snap* the shot the instant the front sight touches the bull's-eye, as he has in a less steady position. After he has learned the principles of correct trigger squeeze with the sandbag rest, he is instructed in the other positions, but for the first half day at least he is not allowed to squeeze the trigger except in the prone position, first with, and then without, the sandbag rest.

(*c*) A great deal of trigger-squeeze exercise is necessary, but it must be carefully watched and coached. Improper trigger-squeeze exercise is worse than none at all.

(*d*) Soldiers should not be allowed to simulate fire until they have been thoroughly instructed in trigger squeeze.

Thereafter, in all drills and field exercises where fire is simulated, they should be cautioned to aim at definite objects and to carry out the correct principles of aiming, squeezing the trigger, and calling the shot.

(2) *Procedure.*—Instruction is individual by the coach-and-pupil method. The aiming targets are like those described for position exercises. The trigger-squeeze exercise is conducted at will in the manner outlined for position exercises.

(3) *Duties of coach.*—In the trigger-squeeze exercise the coach sees that—

(*a*) The sights are blackened.

(*b*) The sling is properly adjusted, is tight enough to give support, and is high up on the arm.

(*c*) The proper position is taken.

(*d*) The slack is taken up promptly, and with a firm initial pressure.

(*e*) The pupil aims. (The coach checks occasionally by means of the aiming device.)

(*f*) The breath is held while aiming. (The coach checks the breathing by watching the back of the pupil.)

(*g*) The trigger is squeezed properly.

(*h*) The pupil calls the shot.

(*i*) The pupil brings the rifle down if the trigger pressure is not completed within 10 seconds after it was first applied.

(*j*) The pupil's eye stays open when the striker falls.

(*k*) The pupil stays relaxed. The coach has the pupil bring the rifle down if he notes tenseness or signs that the pupil is about to flinch (detected by twitching of the eye and the position of the cheek on the thumb or stock).

(*l*) The front sight does not flip. (The coach alines the sight on some object like a blade of grass or a twig.)

■ 47. FOURTH STEP: RAPID FIRE.—*a. General principles.*—(1) The importance of correct rapid-fire training cannot be too strongly emphasized. This is a phase of instruction that is often neglected. Rapid fire is the true test of the good shot. Superiority of fire in battle depends upon fire power, and fire power is the product of accuracy and volume. Rapid fire without accuracy has no value. Accuracy must not be sacrificed for rapidity. Through training, accurate fire becomes more and more rapid, until the ability to fire from 10 to 15 accurate shots per minute is acquired, but no man is permitted to attempt to fire more than 10 shots per minute until he has had long training on the rifle range and has become a seasoned shot. Recruits having target practice for the first time should not be admonished for failing to fire the prescribed number of rounds in the stated time limit. It is advisable to extend the time 10 or 15 seconds when they first begin rapid-fire practice on the range.

(2) All points learned in slow fire are applied in rapid fire. It is especially important that the men understand that the aim and the trigger squeeze are the same as in slow fire. Time is gained by taking position rapidly, by working the bolt rapidly, by reloading the magazine quickly and without fumbling, and by keeping the eye on the target while working the bolt.

(3) The man who looks into the chamber while working the bolt works the bolt slowly. He loses time in finding his target again and often fires on the wrong target. The application in war is apparent; a soldier who takes his eye off an indistinct target to look into the chamber while working the bolt may be unable to relocate his target. The coach must see that the pupil looks at the target constantly, not at the bolt or chamber.

(4) Success in rapid fire is largely a matter of correct form—form in taking positions without lost motion, smooth bolt manipulation, correct manner of breathing, quick sight alinement, decisive taking up of slack, concentration on the sight picture, and continuous, steady, and smooth application of the trigger squeeze.

b. Bolt-operation exercise.—(1) *General.*—(*a*) This exercise is for the purpose of acquiring a smooth and rapid bolt operation. The magazine floor plate, magazine spring, and follower must be removed from the rifle or the follower must be depressed to prevent it from blocking the bolt. A convenient method of doing this is to place an empty cartridge clip edgewise between the follower and the right side of the receiver. Cam surfaces on the bolt should be lubricated to avoid undue wear during the exercise.

(*b*) Practice in bolt operation should be conducted in all positions, and no pupil should be considered proficient until he can operate the bolt at least 20 times in 20 seconds while in the prone position. The first hour of rapid-fire training should be devoted to bolt-operation exercises. Thereafter each pupil should be given additional practice from time to time until he is considered proficient.

(2) *Procedure*.—The exercise is conducted by the coach-and-pupil method. The instructor explains and demonstrates that—

(*a*) The follower is removed or depressed.

(*b*) The correct position is assumed.

(*c*) The bolt handle is gripped firmly by the thumb and first two fingers of the right hand and snapped up, fully back, forward, and fully down with one motion, instead of by four distinct motions as men are inclined to do at first.

(*d*) The eye is kept on the target.

(*e*) The right hand is brought to the small of the stock, the rifle to a horizontal position, and the cheek placed against the thumb or stock each time the bolt is closed.

(*f*) The butt of the rifle is kept against the shoulder.

(*g*) In the prone and sitting position the elbows are kept in the firing position; in the kneeling position the left elbow is kept in position, the right elbow is raised well above the height of the shoulder, so as to depress the muzzle, the elbow is lowered to the firing position just before the bolt reaches its forward position.

(*h*) To provide additional leverage it may be necessary, particularly in the kneeling position, to lower the muzzle and slightly raise the upper part of the body as the bolt is opened.

(*i*) No attempt is made to aim or to squeeze the trigger, but the trigger finger is placed inside the trigger guard each time the bolt is closed.

(*j*) The proper manner of breathing during the exercise is to inhale as the bolt is opened and exhale as the bolt is closed.

(*k*) Emphasis is placed upon the detailed duties of the coach which are listed in (3) below. Exercises should not be continued longer than about 20 seconds at a time. Frequent changes of coach and pupil are necessary to prevent undue fatigue. After requiring the pupils to assume a firing position, the instructor commands: 1. BOLT-OPERATION EXERCISE, READY, 2. EXERCISE, 3. CEASE FIRING, 4. REST.

(3) *Duties of coach.*—In the bolt-operation exercise the coach sees that—

(*a*) The bolt is operated properly and rapidly.

(*b*) The eye is kept on the target.

(*c*) The right hand is brought to the small of the stock, the rifle to a horizontal position, and the cheek placed against the stock each time the bolt is closed.

(*d*) The butt of the rifle is kept against the shoulder.

(*e*) The elbows are kept on the ground in the prone position; between the legs in the sitting position; and the right elbow raised above the height of the shoulder in the kneeling position.

(*f*) The trigger finger is placed in the trigger guard each time the bolt is closed.

(*g*) Breathing is done properly.

(*h*) No attempt is made to aim or press the trigger.

c. Taking positions rapidly.—(1) *Prone position.*—(*a*) *Ready position.*—First assume the prone position and aim at the target as explained in paragraph 45*g*. Mark the position of the elbows and the point on the ground below the butt of the rifle. Rise to the knees and then to the feet without changing the position of either foot. Grasp the rifle with the left hand just below the lower band and place the right hand at the heel of the stock. This is the *ready position*. Being at the READY, there are two methods of taking the prone position rapidly. These methods are described by the numbers for the purpose of instruction in the sequence of the movements. After this sequence is learned the position is taken in one motion.

(*b*) *First method* (fig. 35).

1. Bend both knees to the ground.

2. Place the butt of the rifle on the ground at the point marked.

3. Place the left elbow on the ground.

4. Place the butt of the rifle against the right shoulder with the right hand, at the same time spread the feet apart.

5. Grasp the small of the stock with the right hand and place the elbow on the ground.

6. Adjust the body to place the proper tension on the sling and place the cheek in position against the stock.

FIGURE 25.—First method of taking prone position rapidly.

FIGURE 25.—First method of taking prone position rapidly—Continued.

(c) *Second method* (fig. 26).

1. Throw the right foot well back, bend the left knee as low as possible, and place the butt of the rifle on the ground 4 or 5 inches to the left and slightly in front of the spot where the right elbow is to rest. (The grip on the rifle is retained with both hands.)

①

FIGURE 26.—Second method of taking prone position rapidly.

2. Place the right elbow on the ground.

3. Place the left leg back near the right one, feet apart, and slide well back while lying on the belly.

4. Take the butt of the rifle off the ground and place it against the right shoulder.

5. Lower the left elbow to the ground.

6. Adjust the body to place the proper tension on the sling and place the cheek in position against the stock.

(2) *Sitting position.*—(*a*) To assume the sitting position rapidly, break the fall by placing the right hand on the ground slightly to the right rear of the spot on which to sit.

(*b*) In practicing for range firing, first sit down and aim at the target in the normal sitting position. Then mark the position of the heels and the spot on which to sit. At the

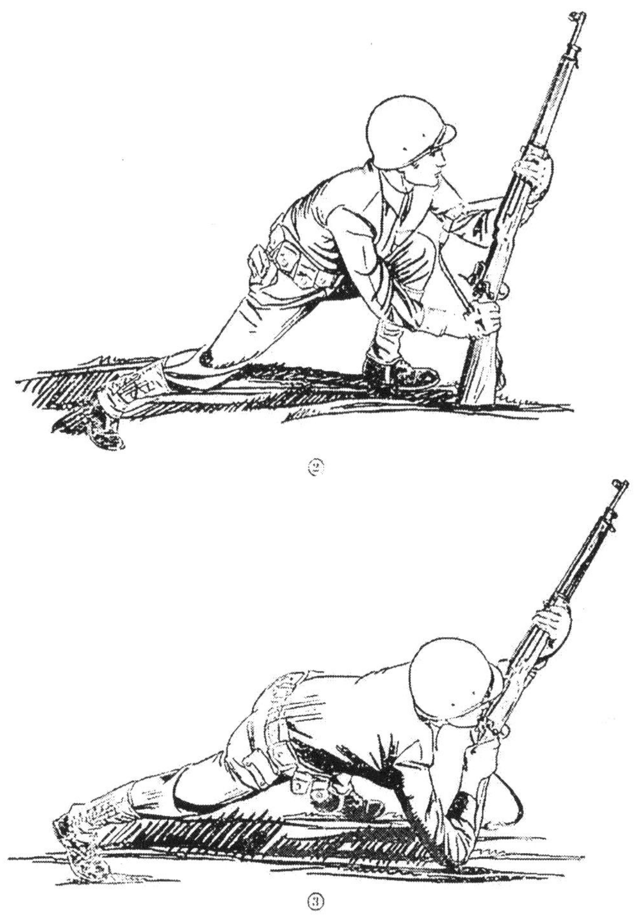

FIGURE 26.—Second method of taking prone position rapidly—Continued.

command READY ON THE FIRING LINE, stand with the heels in the places made for them. As the target appears, sit down on the spot marked, breaking the fall with the right hand.

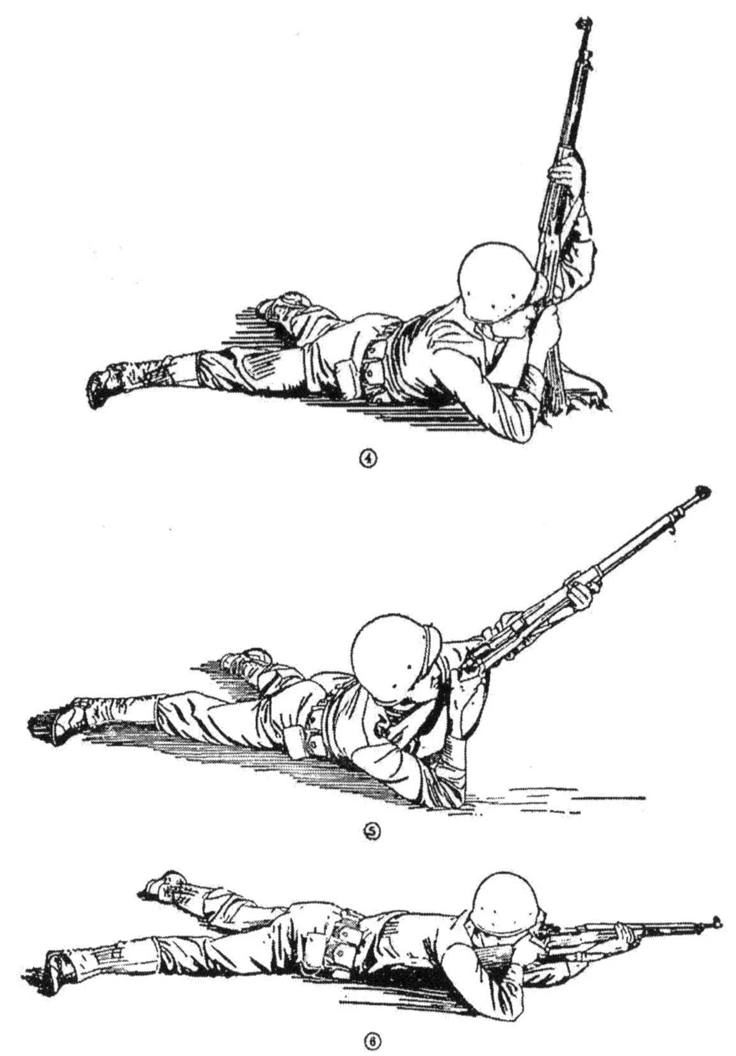

FIGURE 26.—Second method of taking prone position rapidly—Continued.

Place the butt of the rifle on the shoulder with the right hand, grasp the small of the stock with the right hand, and assume the aiming position. (See fig. 22.)

(3) *Kneeling position.*—To assume the kneeling position rapidly, first kneel and aim at the target in the normal

kneeling position. Mark the position of the feet and the right knee. At the command READY ON THE FIRING LINE, stand with the feet in the places marked for them. As the target appears, kneel with the right knee on the spot marked, place the butt of the rifle on the shoulder with the right hand, grasp the small of the stock with the right hand and assume the aiming position. (See fig. 23.)

(4) *Practice required.*—Taking positions rapidly should be practiced at will, using the coach-and-pupil method.

d. Rapid-fire exercise.—(1) *General.*—The instructor assembles his squad or group and explains and demonstrates—

(*a*) The correct method of loading a clip of ammunition into the magazine, using dummy cartridges for the demonstration.

(*b*) The working of the bolt while in the prone and sitting positions.

(*c*) The importance of correct breathing.

(*d*) The time saved by counting the shots of the first clip.

(*e*) The disadvantage of looking into the chamber while working the bolt and into the magazine to see if there are any remaining cartridges.

(*f*) That the follower holds the bolt back when the magazine is empty.

(*g*) That the butt of the rifle is never taken from the shoulder when working the bolt, except for reloading.

(*h*) That a quick but deliberate movement of the hand in taking the clip from the belt and placing it in the clip slot is more effective than a hurried movement.

(2) *Procedure.*—In conducting rapid-fire exercises, the group under instruction is paired off, coach and pupil, and placed on line. Full-size targets are placed at 200 and 300 yards from the men under instruction, and a simple arrangement is made to permit exposure of the target for the prescribed period of time. Rapid-fire exercises may be conducted at shorter ranges using targets proportionately reduced in size. Sights are set to correspond to the range being used. The commands and procedure are exactly the same as in rapid fire on the rifle range except that practice dummy ammunition is used. For example, the pupil stands with sights properly set and blackened, sling adjusted on his arm,

and with two clips of practice dummy cartridges in his belt. The instructor, after announcing the range and the position to be used, commands: 1. WITH DUMMY CARTRIDGES, LOAD, 2. READY ON THE RIGHT, 3. READY ON THE LEFT, 4. READY ON THE FIRING LINE, 5. CEASE FIRING, 6. UNLOAD. At the first command the rifle is loaded and locked. At the fourth command the rifle is unlocked. When the target is exposed the pupil takes position rapidly and simulates firing. He attempts to fire 10 rounds, using proper performance, in the allotted time. Upon completion of the exercise any cartridges remaining in the rifle are removed and the bolt is left open. The sight leaf is immediately laid. During simulated firing the soldier should never take his eye from the target except to reload. He should count his shots as he fires in order to know when the magazine is empty, thus avoiding the loss of time incident to the attempt to shove the bolt forward when the follower blocks it. The exercise is conducted from the standing position to the prone, sitting, and kneeling positions.

e. Duties of coach.—In a rapid-fire exercise the coach sees that—

(1) The sights are blackened.

(2) The gun sling is properly adjusted.

(3) The correct position is taken.

(4) The slack is taken up promptly.

(5) The breath is held while aiming and the firer inhales and exhales while working the bolt.

(6) The trigger is squeezed properly.

(7) The bolt is worked rapidly.

(8) The shots of the first clip are counted.

(9) The eye is kept on the target, the elbows kept in place, and the butt of the rifle kept to the shoulder while working the bolt.

(10) The magazine is reloaded quickly and without fumbling.

■ 48. FIFTH STEP: EFFECT OF WIND; SIGHT CHANGES; USE OF SCORE BOOK.—*a. Wind.*—(1) In firing at 500 yards or under, the effect of the weather conditions, except wind, can be disregarded. The influence of wind on the bullet must be carefully studied.

(2) The horizontal clock system is used in describing the direction of the wind. The firing point is considered the center of the clock. The target is at 12 o'clock. A 3 o'clock wind comes directly from the right. A 6 o'clock wind comes directly from the rear. A 9 o'clock wind comes directly from the left. A wind that is constantly changing its direction back and forth is called a *fishtail wind.*

(3) The force of the wind is described in miles per hour. The force of the wind is estimated by throwing up dry grass, dust, or a bit of light paper and watching how fast it travels, and by observing the danger flags.

In general, a light breeze is a 5 to 8 mile wind; a fairly strong breeze is a 10 to 12 mile wind. Wind blowing 20 miles an hour is very strong.

(4) Wind from either side blows the bullet out of its path. For example, if the wind is coming from the right of the firer (that is, blowing against the right side of the bullet), the bullet will be blown to the left, and will hit left of the bull's-eye. The amount the bullet will be blown from its path depends on the force and direction of the wind and on the distance to the target. Correction for this is made by moving the aiming point toward the wind. This process is called "taking windage." It is accomplished by shifting the body so that the sights of the rifle, instead of pointing at the normal point of aim, are pointed to the right or left of the normal aiming point, depending upon the direction of the wind. The point of impact of the bullet will move in the same direction that the aim is moved. For example, if it is desired to move the hits to the left, the aim must be moved to the left; if it is desired to make the hits strike lower and to the right, the aim must be moved down and to the right.

(5) The amount of windage to allow for the first shot can be determined by one of the methods described in (*a*) and (*b*) below.

(*a*) Refer to the windage diagram for the M1903 rifle in the score book and note the value of the wind in points; then extend the vertical mark on the bottom or top edge of the recording target in the score book, corresponding to the value of the wind; draw a light line tangent to the bottom of the bull's-eye across the target; the intersection of these lines is the point of aim for the first shot, and for successive shots if the point proves correct.

(*b*) By the wind rule method (approximate only), the range (expressed in hundreds of yards) multiplied by the velocity of the wind and divided by 10 equals the number of quarter points to allow for a 3 o'clock or 9 o'clock wind.

Example: At 500 yards the wind is blowing at 15 miles per hour from 3 o'clock; $\frac{5\times 15}{10}$ equals 7.5 quarter points or (approximately) 1¾ points of windage. Refer to the corresponding windage mark on the bottom or top edge of the recording target in the score book for 500 yards; extend this mark up or down the recording sheet; draw a light line tangent to the bottom of the bull's-eye across the sheet; the intersection of these lines is the point of aim on the target for the first shot, and successive shots if the first one is found correct.

(6) If a further correction is necessary after the first shot is fired, it is determined as follows: The shot is plotted on the recording target; from the vertical center line through the bull's-eye the proper correction is noted; and the aiming point for the following shot is shifted in the proper direction by this amount.

(7) As the direction of the wind approaches 12 or 6 o'clock, less and less windage is required. Strong winds from 12 o'clock tend to retard the bullet slightly, and winds from 6 o'clock have a tendency to accelerate it. However, the amount of deviation is so slight that a correction is seldom required.

b. Elevation rule.—(1) Changing the elevation 100 yards at any range will give a change on the target, in inches, equal to the square of the range (expressed in hundreds of yards).

Example: At 200 yards, changing the elevation 100 yards changes the point of strike on the target 4 inches; at 300 yards, 9 inches; at 500 yards, 25 inches. This rule is not exact but is near enough for all practical purposes.

(2) The horizontal marks on the right and left edges of the recording targets in the score book show how much change to make in the elevation at each range. When a change in

elevation is necessary, it is best to refer to these marks before deciding how much change to make.

c. Light.—(1) Light has no effect on the bullet but does affect the aim. The effect of changes of light is very slight with most riflemen. The correction for variations in light does not exceed 25 yards in elevation at any range. The effect of changes of light is not uniform in its effect upon the aim of all riflemen.

(2) As a general rule men unconsciously aim a little lower in a poor light than in a good light and consequently need more elevation when the light is poor. This lowering of the aim is caused by the fact that the outline of the bull's-eye is not distinct in a poor light; therefore men cannot hold as close to the bull's-eye and still be sure of their aim. As a rule poor light exists on dark days when there is a haze in the air, on very bright, warm days, and when the sun is back of the target. The best light for shooting is when the sky is uniformly overcast and there is sufficient light to see the target clearly.

(3) Sunlight from one side has the same effect with most men as wind from that side. This is because the side of the front sight toward the sun is more clearly defined and the tendency is to bring that side under the center of the bull's-eye. Such holding places the bullet on the opposite side of the bull's-eye from the sun. When it is necessary to make an allowance for this, the aiming point is moved toward the sun.

d. Zero of rifle.—(1) The zero of a rifle for each range is the elevation to be set on the rear sight, and the point on the target on which the sights must be alined in order to hit the center of the bull's-eye on a normal day when there is no wind. This zero may not conform to the marks on the scale of the sight leaf and the alinement of the sights on the rifle. The zero of any rifle may differ with different men, owing to the difference in their hold or manner of aim.

(2) Each man must determine the zero of his own rifle for each range. He does this by studying the data which he has written in his score book concerning elevation settings on the rear sight, aiming points, changes in sight settings and aiming points, light, and the direction and velocity of the

wind. The zero of a rifle is best determined on a day when the sky is overcast and there is no wind. Having learned the zero of his rifle, the rifleman computes his elevation for the first shot from this zero and not from the zero marked on the rifle sight unless the two correspond. Likewise, he determines his aiming point from the zero aiming point and not from the normal six o'clock aiming point, unless the two correspond.

e. Shooting up or down hill.—In shooting either up or down hill, less elevation is needed than when shooting on the level. The steeper the hill the less elevation is needed, so that when firing vertically, either up or down, no elevation at all is needed. The slight slopes found occasionally on target ranges have no appreciable effect upon the elevation used and no correction is required.

f. Sight-setting and sight-changing exercises.—In these exercises the instructor uses the full-size A, B, and D targets, with spotters to indicate the position of the hits.

(1) The instructor assembles his squad or group (each pupil having his rifle, score book, and pencil) and conducts the exercise as follows:

(*a*) He points out the graduations on the sight leaf and explains that the line directly opposite the number represents the range; points out the index lines on the slide and explains that in setting the sight for any range these lines must coincide with the graduations for that range on the sight leaf, and that to retain the slide in adjustment the slide catch must be engaged with the proper notch on the edge of the sight leaf.

(*b*) He explains the effect of wind and cautions the class to disregard all atmospheric influences except wind.

(*c*) He shows the pupils how to draw the elevation and windage lines for each range on the recording targets of the score book upon which they are to plot their shots during range firing.

(*d*) He explains the windage diagram in the score book.

(*e*) He explains and demonstrates the methods of determining the amount of windage (aiming point) for the first shot, and how to make corrections for subsequent shots.

(*f*) He explains and illustrates the wind rule and the elevation rule. By asking questions he assures himself that these rules are understood.

(*g*) Having explained the foregoing points to the assembled group, the instructor places the men in pairs, and tests their ability to set the sights for the first shot by use of the wind rule or the windage diagram for the M1903 rifle in the score book. Every time the sights are set each man examines the sight of the man paired with him. The instructor calls for reports of several of the sight settings examined.

(*h*) The instructor tests the ability of the group to change sights intelligently after the first shot by reference to the horizontal and vertical lines on the model target in the score book.

(2) The following examples of windage exercises are given:

(*a*) "Use the windage diagram for the M1903 rifle for all problems. You are at 500 yards and estimate the wind to be 8 miles per hour from 3 o'clock; plot on the recording target in your score book the aiming point for the first shot. Jones, where has Smith placed his aiming point? Each man whose teammate did not place his aiming point to the right of the bull's-eye at the intersection of a horizontal line drawn tangent to the bottom of the bull's-eye and the vertical line indicating 1 point windage, hold up his hand." The instructor, by questions, explanations, and illustrations assists the men who have made mistakes.

(*b*) "You are at 300 yards and estimate the wind to be 7 miles per hour from 9 o'clock. Plot your aiming point on the recording target in your score book for the first shot. Evans, where has Little plotted his aiming point? Each man whose teammate has not plotted his aiming point to the left of the bull's-eye at the intersection of the 1/2 point vertical line and a horizontal line tangent to the bull's-eye, hold up his hand. Suppose you fired and the spotter marked the hit here (placing a spotter in the 4 space near the bull's-eye at 3 o'clock) and you were sure your hold and trigger squeeze were perfect; plot your aiming point to bring the next shot in the vertical center line of the bull's-eye. Johnson, where has Williams plotted his aiming point? Each man whose teammate has not changed his aiming point for the next shot to the left on the 1 point line, hold up his hand." The instructor assists those men who have decided errors by further explanation and illustrations.

(3) The instructor gives a number of examples with the wind at different angles and velocities and at the various rangers until the class thoroughly understands the methods of determining the amount of windage to take for the first shot and corrections, if necessary, in aiming points for subsequent shots.

(4) Following the instructions in taking windage, the instructor puts the class through similar exercises which require changes in elevation.

(5) Elevation exercises such as the following may be used:

(*a*) "You are at 300 yards, set your sights for 300 yards. There is no wind. Walsh, where is your aiming point for the first shot?" The instructor explains that since there is no wind the aim should be normal. "You fire and at the instant the shot went off you noticed that your sights were alined a little below and a little to the right of your zero aiming point. When the shot is marked you see that it struck a little to the right and 6 inches below (placing a spotter in the 4 space at 5 o'clock). You fire one more shot, and this time the aim was correct at the instant the rifle went off. This time the shot is marked here (placing a spotter about 5 inches in the 4 space just below the bottom edge of the bull's-eye). Allen, what are you going to do now? Lowe, what are you going to do? Both shots hit about the same distance below the bull's-eye. Your trigger squeeze was perfect because you were able to call your shots close to where they actually hit. The variation in aim was very slight, therefore you are justified in making a correction in elevation to raise the next shot to hit the center of the bull's-eye. Set your sights accordingly. Backell, what does Moore's sight setting read? Each man whose teammate does not have his sights set at 400 yards, hold up his hand. Ellis, why is a raise of 100 yards in elevation required to bring the next shot in the center of the bull's-eye?" The instructor by explanations and illustrations assists those men who do not understand the procedure in arriving at the correct result.

(b) A number of examples are given using the elevation marks on the recording sheet of the score book. He emphasizes that this method should always be used when the score book is available.

(6) The instructor gives a number of examples which require changes in both windage (aiming point) and elevation until the principles of aiming point and elevation changes are well understood.

(7) Other sight-setting exercises such as the following are useful:

(a) "Set your sights at 500 yards; and place your aiming point on the 1 point left windage line. Suppose you fire four shots hitting here (place four spotters in the bull's-eye), and your fifth shot here (place spotter on 3 space at 11 o'clock). Jones, what are you going to do now? Jenkins, what are you going to do? You should not do anything to the sight. It is practically certain that you squeezed the trigger improperly and flinched. Not even a sudden and violent change in the weather could make that much difference. Don't try to correct your faults by changing the sights around."

(b) "For your first score in rapid fire at 200 yards you have set your sight at the same elevation and windage that you used in zeroing the rifle on the rapid fire targets. Suppose this to be 200 yards elevation and your group goes here (putting 10 spotters low and to the left). Set your sight to bring the next score into the figure. Miller, what does Wright's sight read?"

(8) A group in rapid fire should strike the same place as in slow fire. When rapid-fire groups vary in position from slow-fire groups the difference is caused by imperfect trigger squeeze in rapid fire, which causes these groups to be more scattered. Men should try to squeeze the trigger so that the rapid-fire sight settings will be the same. But if there is a constant variation in the two sight settings, each man should note it in his score book and set his sight in rapid fire so as to make the groups count as much as possible. Groups that are scattered all over the target cannot be corrected by changing the sight.

g. Use of score book.—(1) Each man must keep a score book in which he records not only the value of the hits but the location of each hit, the sight setting and sight changes, the force and direction of the wind, the kind of light, the hour, the date, and such other data as may be of use in the future. Spaces for these notes are provided on the score sheets of the score book.

(2) The use of the score book on the range is important for the following reasons:

(*a*) The plotting of the shots shows the firer the location of his group.

(*b*) The windage diagram indicates the windage (aiming point) to use for the first shot. The extension of the vertical and horizontal marks on the recording targets of the score book shows the change in elevation and aiming point (windage) necessary to place the group in the center of the target.

(*c*) Plotting the shots and recording the data as to light and wind help the soldier to learn the zero of his rifle.

(*d*) The data written down as to sight settings, aiming points and weather conditions while firing at any range are of great assistance in setting the sight correctly and determining the aiming point when again firing at that range. Where a number of scores have been fired and recorded, the firer should get his sight settings and determine his aiming point from previous scores fired on days that were similar as to light and wind.

(3) The score book is kept personally by the man firing. The coach assists him when necessary to decide what to enter, but the coach neither plots the shots nor enters any data.

h. Score-book exercises.—The squad or larger group is assembled in front of a full-size B target, each man with score book, pencil, and rifle. The class is divided into pairs. Each man acts as a coach for the other man of his pair.

(1) The instructor states the light and weather conditions and the range. He then indicates 10 successive shots on the target by means of a spotter and requires each man to plot each shot as it is indicated, write down the data given from time to time, and make the actual sight settings and corrections on his rifle, and plot the aiming points on the record-

ing target. Weather and light conditions assumed by the instructor and changes announced during the exercise should be those that are likely to occur on the rifle range.

(2) The pupils are told by the instructor to open their score books at the first blank B recording target for the M1903 rifle and plot the shots and write in the data as given to them. They are further instructed to write lightly so that the entries may be erased and the page used again. The example given should be substantially as follows:

(*a*) "You are getting ready to fire a slow-fire score at 500 yards. There is bright sunlight. The wind varies from 8 to 12 miles an hour from 1 to 3 o'clock. When you are in position and ready to fire the first shot, the wind seems steady from 3 o'clock and is blowing about 8 miles an hour. Write in your data, set your sight, and plot your aiming point. Jones, where has Robinson set his sight, and where has he plotted his aiming point? Smith, where has Sharp plotted his aiming point? The sights should be set at 500 yards elevation. The windage called for is ¾ point right, therefore the aiming point (windage) should be to the right of the bottom of the bull's-eye, at a point midway between the ½ and 1 point windage lines.

(*b*) "You fire your first shot, your performance was excellent and you called your shot a five a little above the center of the bull's-eye, the spotter marks it (a close four at 7 o'clock). Decide what you are going to do and make the necessary change in elevation and aiming point to conform to your decision. Dodd, what changes has Snyder made? Since your performance was perfect and the strike of the bullet was about 5 inches to the left and 11 inches below the center of the bull's-eye, corrections should be made in elevation and aiming point for the next shot. The elevation on the sights should be raised to 600 yards, and the aiming point shifted to the right ¼ point and lowered 10 inches. Hence the new aiming point will be at the intersection of the 1 point right windage line and the lower 75 yard elevation (horizontal) line." The instructor explains the reason for raising the sights 100 yards and lowering the aiming point approximately 10 inches, to the 75-yard elevation line. He points out that this procedure is better than not raising the elevation on the sight and raising the aiming point the required 10 inches, because raising the aiming point would partially obscure the bull's-eye, and since it is used as the main reference object in aiming, corrections in elevation should be made by using a combination of change in sight setting and aiming point so that all of the bull's-eye is always visible.

(*c*) "Your second shot hit here (spotter near center of bull's-eye); your third shot hit here (spotter in bull's-eye near the top); your fourth shot hit here (spotter in bull's-eye near the bottom); your fifth shot hit here (close 4 at 9 o'clock). The wind seems to be a little stronger, but you are not sure. Your performance and call were all right. Brown what are you going to do? Half the correction indicated by the windage lines on the recording target of the score book is correct. Plot your hit and change your aiming point accordingly. The recording target indicates full correction requires ½ point right, half correction would be ¼ point right, therefore the aiming point for the next shot should be moved to the right to a point half way between the 1 point and 1½ point windage lines, that is, you are now using 1¼ point right windage. Your sixth shot hits here (a bull's-eye near the edge at 3 o'clock). Baker, what are you going to do now? It appears that the fourth and fifth shots were caused by faulty trigger squeeze, so change your aiming point back to where it was before (the intersection of the 1 point right line and the lower 75-yard elevation line).

(*d*) "Before you fire your seventh shot you notice that the wind has shifted to about 1 o'clock but is still blowing at the same velocity (8 miles per hour). Gallman, where is Rozinski's aiming point now? Huntley, what does the windage diagram show about a 1 o'clock wind? You need only about half as much windage as for a 3 o'clock wind. Your aiming point should therefore be moved *to the left* on the ½ point *right* windage line. Place it there.

(*e*) "Your next shot hits here (a wide 4 at 6 o'clock). Clement, what corrections has Ferenc made for his eight shot? You should have made no change in your sights or your aiming point. Your windage is apparently correct and there has been no change in the conditions. Your low shot was caused by poor performance, possibly poor sight alinement but more likely poor trigger squeeze. Don't try to correct poor performance by changing sights or aiming point. Your eighth shot hits here (deep in the bull's-eye at 9 o'clock); your ninth shot here (bull's-eye, deep in at 4 o'clock); your tenth shot hits here (bull's-eye at 1 o'clock). Write in your notes and exchange books with your teammate. Coffman, has Doucet plotted all of the shots correctly? Read the notes he has written in his book."

(3) The instructor corrects errors and mistaken ideas, and notes those men who need additional instruction.

■ 49. SIXTH STEP: EXAMINATION OF MEN BEFORE STARTING RANGE PRACTICE.—(The answers given to the questions in this paragraph are guides only. Men should answer the questions in their own words.)

Q. What is this (showing the cut-out circle of a piece of cardboard)? *A.* A circle or aperture representing the peep sight.

Q. Where is the center of it? *A.* Here (pointing to the center).

Q. What does this represent (showing an improvised enlarged front sight made of cardboard)? *A.* It represents the front sight of the rifle.

Q. Where should the top of the front sight be when it is correctly aligned in the peep sight? *A.* The top of the front sight should be in the center of the peep sight. (By use of the movable sights the instructor requires the pupil to adjust the front sight in proper alinement with the peep sight.)

Q. What is this (showing a movable bull's-eye)? *A.* The bull's-eye.

Q. In aiming, is the bull's-eye in the center of the peep sight? *A.* No; the bottom edge of it is in the center.

Q. Why? *A.* Because the top of the front sight is in the center and just touches the bottom of the bull's-eye.

Q. Should the front sight be held up into the bottom of the bull's-eye? *A.* No; it just touches the bottom edge of the bull's-eye, so that all of the bull's-eye can still be clearly seen.

Q. Why shouldn't the front sight be held up into the bull's-eye? *A.* Because the front sight would blend with the bull's-eye and its position could not be accurately determined each time the aim is taken. Furthermore the top of the front sight would not be clearly defined and could not be accurately centered in the peep. (The instructor requires the pupil to adjust the sights and bull's-eye to show the correct aim.)

Q. What is this (indicating sighting bar)? *A.* Sighting bar.

Q. What is it for? *A.* To teach the correct sight alinement.

Q. Why is it better than a rifle for this purpose? *A.* Because the sights on the sighting bar are much larger and slight errors can be more easily seen and pointed out.

Q. What does this represent? *A.* The front sight.

Q. And this? *A.* The rear sight.

Q. What is this? *A.* The eyepiece.

Q. What is the eyepiece for? *A.* To cause me to place my eye in such a position as to see the sights in the same alinement as that set by the coach.

Q. Is there any eyepiece on the rifle? *A.* No; I learn by the sighting bar how the sights look when properly alined, and I must hold my head so as to see the sights the same way when aiming a rifle.

Q. How do you hold your head steadily in this position when aiming a rifle? *A.* By resting my cheek firmly on the right thumb or against the side of the stock.

Q. Where do you focus your eye when aiming a rifle? *A.* On the bull's-eye.

Q. What do you understand by the term "correct sight picture?" *A.* It means that all the elements of the aim are in proper relationship to each other.

Q. Tell me what is wrong with these sight pictures. (The instructor now adjusts the sights of the bar, making various slight errors; first, to show the correct and incorrect adjustments of the sights, and then, with the sights properly adjusted, he sights on the small bull's-eye to demonstrate correct and incorrect adjustments, requiring the man to point out any errors.)

Q. Now, take this sighting bar and adjust the sights properly. (Verified by the instructor.)

Q. Now that the sights are properly adjusted, have the small bull's-eye moved until the sights are properly aimed at it.

Q. How do you breathe while aiming? *A*. After I get my sights lined up on the bull's-eye, I draw in a little more than an ordinary breath, then let out a little until I have a relaxed, easy feeling in the lungs, and hold the remainder while aiming and squeezing the trigger.

Q. Take the prone position, aim, and simulate firing a shot at that mark. (The instructor must assure himself that the man knows how to hold his breath properly while aiming. Many men have great difficulty in learning to do this correctly.)

Q. What is this? *A*. An aiming device.

Q. What is it used for? *A*. To show the instructor how a man is aiming.

Q. Now, I will take this rifle, and with the sandbag rest to hold the rifle steady I will aim at the bull's-eye. You watch the sights through the aiming device and tell me when my aim is right and when it is wrong. When it is wrong, tell me what the error is. (The instructor now aims so as to illustrate the common faults and the man must observe and point them out.)

Q. I will now simulate firing at a bull's-eye a few times. You watch through the aiming device and call where the shots would have hit.

Q. Now, take this rifle and, using the sand bag rest, aim at the bull's-eye. I will watch you through the aiming device. (The instructor satisfies himself that the man understands sighting and aiming, and requires him to simulate firing a few times and to call his shots.)

Q. I will take the rifle and assume the kneeling, sitting, and prone positions, and you will tell me whether the position is correct or incorrect. (The gun sling is adjusted in all these tests.)

Q. Now you take this rifle and show me the kneeling, sitting, and prone positions and the prone position with the sand bag rest.

Q. Now show me how you take the sitting and prone positions rapidly from a standing position.

Q. When do you take up the slack in the trigger? *A.* When I have held my breath and the sight picture is approximately correct.

Q. When do you start the trigger squeeze? *A.* After I have taken up the slack and checked the sight picture.

Q. How do you squeeze the trigger? *A.* By pressing the trigger straight to the rear so steadily and smoothly that I am not aware of any increase of pressure and I do not know when the rifle will go off.

Q. On what do you check and concentrate while you are squeezing the trigger? *A.* On the sight picture to see that it is correct.

Q. If the sights are slightly out of alinement with the aiming point, what do you do? *A.* I hold the pressure I have on the trigger and only resume the increase of pressure when the sights become lined upon the aiming point again.

Q. If you do this, can your shot be a bad one? *A.* No.

Q. Why? *A.* Because I do not know when the rifle is going off and therefore can't flinch. Moreover, the sights will always be lined up with the aiming point when the rifle goes off, because I only increase the pressure on the trigger when they are properly alined.

Q. Does it take a long time to squeeze the trigger this way? *A.* No. The method of squeezing the trigger is slow at first but rapidity is developed by practice.

Q. How do you squeeze the trigger in rapid fire? *A.* The same way as in slow fire; with a steady, continuous increase of pressure so as not to know when the rifle will fire.

Q. How do you save time in rapid fire so as not to be compelled to hurry in aiming and squeezing the trigger? *A.* I save time by taking the position rapidly, working the bolt rapidly, by keeping my eye on the target while working the bolt, by counting the shots of the first clip, and by making positive and quick movements in reloading.

Q. How does keeping your eye on the target help you to save time? *A.* A man who looks into the chamber while working the bolt loses time because he works the bolt slowly in order to see the cartridge enter the barrel. He also loses time because he has to relocate his target after every shot.

Q. What other mistake often results from "chamber gazing" in rapid fire? *A*. Firing on the wrong target.

Q. Show me how you work the bolt in rapid fire, prone, sitting, and kneeling.

Q. Now show me how you load a clip into the magazine.

Q. Is it important to get into the correct position before beginning to shoot in rapid fire? *A*. Yes; even though it takes more time, I should always get into the correct position before beginning to shoot.

Q. What is meant by "calling the shot"? *A*. To say where you think the bullet hit as soon as you shoot and before the shot is marked.

Q. How can you do this? *A*. By noting exactly where the sights point when the rifle goes off.

Q. If a man cannot call his shot properly, what does it usually indicate? *A*. That he did not squeeze the trigger properly and did not know where the sights were pointed when the rifle went off.

Q. What is this? *A*. A score book.

Q. What are these lines for (indicating the horizontal lines drawn in the recording target of the score book) ? *A*. To show the amount of elevation necessary to bring the shot to the middle line.

Q. If a shot hits here (indicating), what change in your sight would you make to bring the next shot to the center of the bull's-eye?

Q. What effect does changing your aiming point to the right or left have on the shot? *A*. It moves the shot in the same direction in which the aiming point is moved.

Q. If you want to make a shot hit higher, what do you do? *A*. I increase my elevation.

Q. If you want to make your shots hit more to the right, what do you do? *A*. I move my aiming point to the right.

Q. How do you determine the amount of windage to take for the first shot? *A*. First I estimate the velocity and direction of the wind. Next I refer to the windage diagram (of the M1903 rifle) for that range and find the value of that wind *in points*. Then I extend the vertical mark on the recording target of the score book that corresponds to the value of the wind in points. The intersection of this line with a line drawn tangent to the bottom of the bull's-eye gives me my aiming point (windage) for the first shot.

Q. How do you determine the direction of the wind? *A.* I face the target and consider myself as occupying the center of the dial of a horizontal clock. I then note the direction of the wind in reference to the numbers on the dial.

Q. How do you estimate the velocity of the wind? *A.* By estimating how fast smoke, dust, light grass, etc., passes a point.

Q. I will place this spotter on this target (full size 500-yard target) to represent a shot properly fired by you at 500 yards with zero windage and sight set at 500 yards. Take your rifle and move your sight to bring the next shot to the center of the bull's-eye. (Instructor now tests in various ways the man's ability to make proper sight and aiming point corrections.)

Q. What are the three principal uses of the score book? *A.* To show where the shot group is located; to indicate how much change in elevation and windage (aiming point) is necessary to move a shot or group of shots to the center of the target; and to make a record of the sight settings and aiming points of my rifle for the different ranges under various weather conditions so that I will know where to set my sight and where to aim when starting to shoot at each range under different weather conditions.

Q. Tell me what effect different light and weather conditions have on a man's shooting?

Q. In firing at ranges up to and including 500 yards, what is the only weather condition for which you make corrections? *A.* Wind.

Q. What three things do you do in cleaning a rifle after it has been fired? *A.* I first remove the powder fouling from the bore. I then dry the bore thoroughly. After this is done I protect the bore from rust with a thin coating of light preservative lubricating oil.

Q. How do you remove the powder fouling from the bore? *A.* By swabbing it thoroughly with rifle bore cleaner. If this is not available I use hot, soapy water or plain hot water.

Q. How do you dry the bore? *A.* By running clean patches through the bore until it is thoroughly dry.

Q. How do you protect the bore from rust? *A.* By swabbing it thoroughly with a cleaning patch saturated with light preservative lubricating oil which is issued for this purpose.

Section III

QUALIFICATION COURSES

■ 50. General.—See AR 775-10 for information as to who will fire the several courses, individual classification, qualification, ammunition allowances, etc.

■ 51. Course A.—*a. Instruction practice.*

Table I.—*Slow fire*

Range (yards)	Time	Shots	Target	Position	Sling
200	No limit	5	A	Prone, sandbag optional	Loop.
300	No limit	5	A	Prone, sandbag optional	Loop.
500	No limit	5	B	Prone, sandbag optional	Loop.

Table II.—*Slow fire*
(To be fired twice)

Range (yards)	Time	Shots	Target	Position	Sling
200	No limit	5	A	Prone	Loop.
300	No limit	5	A	Prone	Loop.
500	No limit	10	B	Prone	Loop.

Table III.—*Slow fire*

Range (yards)	Time	Shots	Target	Position	Sling
200	No limit	10	A	Sitting	Loop.
200	No limit	10	A	Kneeling	Loop.
200	No limit	10	A	Standing	Hasty.

TABLE IV.—*Rapid fire*

(To be fired twice)

Range (yards)	Time (seconds)	Shots	Target	Position	Sling
200	60	[1] 5	D	Sitting from standing	Loop.
200	60	[1] 5	D	Kneeling from standing	Loop.
300	70	[1] 5	D	Prone from standing	Loop.

[1] See paragraph 56*d*.

TABLE V.—*Rapid fire*

Range (yards)	Time (seconds)	Shots	Target	Position	Sling
200	60	10	D	Sitting from standing	Loop.
200	60	10	D	Kneeling from standing	Loop.
300	70	10	D	Prone from standing	Loop.

b. Record practice.

TABLE VI.—*Slow fire*

Range (yards)	Time	Shots	Target	Position	Sling
200	No limit	10	A	Standing	Hasty.
300	No limit	10	A	5 kneeling; 5 sitting	Loop.
500	No limit	10	B	Prone	Loop.

TABLE VII.—*Rapid fire*

Range (yards)	Time (seconds)	Shots	Target	Position	Sling
200	60	10	D	Kneeling from standing	Loop.
200	60	10	D	Sitting from standing	Loop.

TABLE VIII.—*Rapid fire*

Range (yards)	Time (seconds)	Shots	Target	Position	Sling
300	70	10	D	Prone from standing	Loop.

■ 52. COURSE B.—*a. Instruction practice.*

TABLE I.—*Slow fire*

Range (yards)	Time (seconds)	Shots	Target	Position	Sling
200	No limit	5	A	Prone, sand bag optional	Loop.
300	No limit	5	A	Prone, sand bag optional	Loop.

TABLE II.—*Slow fire*

(To be fired twice)

Range (yards)	Time	Shots	Target	Position	Sling
200	No limit	10	A	Sitting	Loop.
200	No limit	10	A	Kneeling	Loop.
200	No limit	10	A	Standing	Hasty.

TABLE III.—*Rapid fire*

(Parts of this table to be fired twice)

Range (yards)	Time (seconds)	Shots	Target	Position	Sling
200	60	[1] 5	D	Sitting from standing	Loop.
200	60	[1] 5	D	Kneeling from standing	Loop.
300	70	[1] 5	D	Prone from standing	Loop.

[1] See paragraph 56*d*.

TABLE IV.—*Rapid fire*

Range (yards)	Time (seconds)	Shots	Target	Position	Sling
200	60	10	D	Kneeling from standing	Loop.
200	60	10	D	Sitting from standing	Loop.
300	70	10	D	Prone from standing	Loop.

b. Record practice.

TABLE V.—*Slow fire*

Range (yards)	Time	Shots	Target	Position	Sling
200	No limit	10	A	Standing	Hasty.
300	No limit	10	A	5 kneeling; 5 standing	Loop.

TABLE VI.—*Rapid fire*

Range (yards)	Time (seconds)	Shots	Target	Position	Sling
300	70	10	D	Prone from standing	Loop.

TABLE VII.—*Rapid fire*

Range (yards)	Time (seconds)	Shots	Target	Position	Sling
200	60	10	D	Sitting from standing	Loop.
200	60	10	D	Kneeling from standing	Loop.

■ 53. COURSE C.—*a. Instruction practice.*

TABLE I.—*Slow fire*

(To be fired twice)

Range (yards)	Time	Shots	Target	Position	Sling
200	No limit	5	A	Prone, sandbag optional	Loop.
200	No limit	5	A	Prone	Loop.
200	No limit	5	A	Sitting	Loop.
200	No limit	5	A	Kneeling	Loop.
200	No limit	5	A	Standing	Hasty.

TABLE II.—*Rapid fire*

(Parts of this table to be fired twice)

Range (yards)	Time (seconds)	Shots	Target	Position	Sling
200	60	[1] 5	D	Sitting from standing	Loop.
200	60	[1] 5	D	Kneeling from standing	Loop.
200	60	[1] 5	D	Prone from standing	Loop.

[1] See paragraph 56*d*.

b. Record practice.

TABLE III.—*Slow fire*

Range (yards)	Time	Shots	Target	Position	Sling
200	No limit	10	A	5 sitting; 5 kneeling	Loop.
200	No limit	10	A	Standing	Hasty.

TABLE IV.—*Rapid fire*

Range (yards)	Time (seconds)	Shots	Target	Position	Sling
200	60	10	D	Kneeling from standing	Loop.
200	60	10	D	Sitting from standing	Loop.

■ 54. COURSE D.—*a. Instruction practice.*

TABLE I.—*Slow fire*

Range (inches)	Time	Shots	Target	Position	Sling
1,000	No limit	10	A, 1,000 inch	Prone	Loop.
1,000	No limit	10	A, 1,000 inch	Prone	Loop.
1,000	No limit	10	A, 1,000 inch	Sitting	Loop.
1,000	No limit	10	A, 1,000 inch	Kneeling	Loop.
1,000	No limit	10	A, 1,000 inch	Standing	Hasty.

TABLE II.—*Rapid fire*

Range (inches)	Time (seconds)	Shots	Target	Position	Sling
1,000	60	[1] 5	D, 1,000 inch	Prone from standing	Loop.
1,000	60	[1] 5	D, 1,000 inch	Sitting from standing	Loop.
1,000	60	[1] 5	D, 1,000 inch	Kneeling from standing	Loop

[1] See paragraph 56*d*.

b. Record practice.

TABLE III.—*Slow fire*

Range (inches)	Time	Shots	Target	Position	Sling
1,000	No limit	10	A, 1,000 inch	5 kneeling; 5 sitting	Loop.
1,000	No limit	10	A, 1,000 inch	Standing	Hasty.

TABLE IV.—*Rapid fire*

Range (inches)	Time (seconds)	Shots	Target	Position	Sling
1,000	60	10	D, 1,000 inch	Kneeling from standing	Loop.
1,000	60	10	D, 1,000 inch	Sitting from standing	Loop.

SECTION IV

RANGE PRACTICE

■ 55. GENERAL.—*a. Phases.*—Range practice is initiated immediately after completion of the preparatory training. Range practice is divided into two parts—instruction practice and record practice.

b. Sequence of practice.—The practice season opens with instruction practice. Each person will complete instruction practice before he proceeds with record practice. When record practice is once begun by an individual it is completed before any other practice is permitted. As a rule, record practice will not be fired by any rifleman on the same day that he fires any part of instruction practice. However, when the time allotted to range practice is very limited, the officer in charge of firing may authorize record firing on the same day. Instruction practice and record practice will not be conducted simultaneously except on ranges where the firing points are in echelon or where the two types of practice are conducted on different parts of the same range.

c. Range personnel.—(1) *Officer in charge of firing.*—An officer in charge of firing is designated by the responsible commander. It is desirable that he be the senior officer of the largest organization occupying the range. The officer in charge of firing, or his deputy, is present during all firing and is in charge of the practice and safety precautions on the range.

(2) *Range officer.*—The range officer is appointed by the post commander and is responsible to him for maintaining and assigning ranges, for designating danger zones, and for closing roads leading into danger zones. The range officer makes timely arrangements for material and labor to place the ranges in proper condition for range practice. He directs and supervises all necessary repairs to shelters, butts, targets, firing points, and telephone lines. He provides for the safety of the markers, and when necessary he provides range guards and instructs them in the methods to be used for the protection of life and property within the danger area. He assists the officer in charge of firing by

using the means necessary to provide efficient service from the maintenance personnel of the ranges.

(3) *Range noncommissioned officer.*—A noncommissioned officer and such assistants as the post commander may deem necessary are detailed permanently during the range practice season as assistants to the range officer. The range noncommissioned officer is responsible to the range officer for maintaining the target and pit equipment in a serviceable condition; for seeing that the desired targets are ready for use at the appointed time; and for providing all target and pit details with the proper flags, marking disks, pasters, and spotters.

(4) *Pit details.*—Commanders of organizations firing provide such details of officers, noncommissioned officers, and privates as may be necessary to supervise, operate, mark, and score the targets used by their respective organizations (see par. 57*c*).

d. Uniform.—The uniform to be worn during instruction practice and record practice is prescribed by the commanding officer.

e. Pads.—Men should be required to wear pads on the shoulder and, if the ground is hard, on the elbows for the first 3 or 4 days at least. A pad can easily be improvised by putting a pair of woolen socks under the shirt so as to protect the shoulder and the upper muscles of the arm. After a few days of firing the muscles become hardened and pads are not essential (see par. 57*h*(10)).

f. Cartridge belt.—The cartridge belt is worn during instruction practice and record firing.

g. Safety precautions.—(1) Bolts of rifles in rear or on the flanks of the firing line, except those in use during supervised preparatory training, are kept open.

(2) Rifles are not loaded in rear of the firing line.

(3) Loaded rifles are always pointed in the general direction of the targets.

(4) On open 1,000-inch ranges having no danger area behind the backstop, all loading and unloading is executed with the muzzle directed toward the target, and in rapid fire rifles are not unlocked until the firer is in the prescribed firing position.

(5) Ammunition at the firing point should be protected from the direct rays of the sun.

(6) At the completion of each day's firing, and preferably before leaving the range, organization commanders should cause all rifles and personnel to be inspected to insure that rifles are clear of ammunition and that no ammunition is in the possession of individuals.

(7) See AR 45–30 for regulations covering report of accident involving ordnance matériel.

■ 56. INSTRUCTION PRACTICE.—*a. General.*—Instruction practice represents the application with service ammunition of the principles taught in the preparatory training. The instruction practice outlined for each course described in paragraphs 51 to 54, inclusive, is designed to serve as a guide only. The general plan of beginning the practice at all ranges with the sandbag rest should, however, be followed. Within authorized ammunition allowances the number of shots to be fired at each range is discretionary with the organization commander. The amount of instruction practice is not limited to that outlined in the tables. Such additional practice as time and ammunition allowances permit should be given.

b. Organization of firing line.—(1) *General.*—The firing line is organized so as to insure safe and orderly conduct, to provide the maximum training and instruction, to keep all men busy all the time, and to facilitate supervision of the firing by the officer in charge of firing and his assistants.

(2) *Procedure.*—(*a*) If space permits, two teams, each consisting of a pupil and his coach, are assigned to a target. One team works on the right and the other team on the left of the stake indicating the number of the target, and the firing point. The pupil on the right, assisted by his coach, fires his instruction practice first. During this firing the pupil of the team on the left "dry shoots." When the firer of the right team completes his instruction practice, the "dry shooter" of the team on the left fires his instruction practice from the same position he used while "dry shooting." During this firing the men of the team on the right exchange places, that is, the coach becomes the pupil and the pupil the coach, and the pupil "dry shoots." The teams and the

individuals within the teams thus alternate firing, dry shooting, and coaching until all men of both teams have completed their practice firing.

(*b*) If sufficient targets are not available to provide one for each pair of teams, the remaining teams are given "dry shooting" practice on the flanks or at a point not less than 50 yards in rear of the firing lines while they await their turn to fire. Safety precautions will be strictly enforced during this practice.

(3) *Application of rules.*—The above organization and procedure are applicable to both slow and rapid fire.

c. Slow fire.—The first few shots fired on the range by beginners will be slow fire from the prone position with the sandbag rest; following this, slow fire from other positions is conducted. The sandbag rest is used at the beginning of the course, not to teach steadiness of hold but to facilitate instruction in the proper method of squeezing the trigger. The sandbag assures such a steady hold that the temptation of the beginner to snap in his shot at the instant the sight touches or drifts past the bull's-eye, which is the cause of nearly all poor shooting, is eliminated. With the sandbag rest the sights can be held fixed at the bottom of the bull's-eye while the firer squeezes the trigger with such a steady pressure that he does not know exactly when the rifle will fire, which is the basis of all good shooting. The habit of the correct trigger squeeze once acquired by firing with a sandbag rest will in all probability be retained while firing prone and in the more unsteady positions—sitting, kneeling, and standing.

d. Rapid fire.—(1) During rapid fire the tendency to jerk the trigger, and consequently to flinch, is very strong. This tendency must be corrected before it becomes a fixed habit.

(2) The tendency to flinch is eliminated by using clips in which half the cartridges are range dummies. The dummy and live cartridges are put into the clips by the coach in such a way that the pupil cannot know which will fire and which will not. Then, if he is not squeezing the trigger with a steady pressure he will flinch or shove his shoulder forward to meet the shock, even when there is a dummy in the chamber and no shock occurs. The flinch is then apparent to the

coach and to everyone in the vicinity, including the man doing the flinching. The result is that he makes a determined effort to squeeze the trigger with a steady pressure for all shots, so as not to appear foolish both to the observers and to himself. During this kind of practice the coach must watch the firer closely to see that he does not look into the chamber in an attempt to see which cartridges are loaded and which are dummies. If he is allowed to look into the chamber while working the bolt, the value of the practice is lost and a very bad shooting habit is acquired.

(3) Range dummy cartridges (fig. 13) are similar in appearance to service ammunition. They are issued for use on the rifle range with loaded cartridges. *Any other use of range dummy cartridges is prohibited.* Practice dummy cartridges of such shape and color as to be readily distinguished from the service cartridge are used in all other exercises that require the use of dummy cartridges.

(4) It is advisable to have each order, when it comes to the firing point, simulate a score of rapid fire, using dummy cartridges or having the cut-off turned down.

e. Coaching.—(1) *General.*—During instruction practice the soldier works under the supervision of a coach. This does not mean that each man must have an experienced shot beside him. Any man of intelligence who has been properly instructed in the preparatory work and who has been given instruction in coaching methods can be used with good results and should be used when more experienced shots are not available. It is good practice to have expert coaches in charge of one or more targets, usually on a flank, to which particularly difficult pupils are sent for special coaching. Great patience should be exercised by the coach so as not to excite or confuse the firer.

(2) *Position of coach* (fig. 27).—On the firing line the coach takes a position similar to that of the man who is firing—prone, sitting, kneeling, or standing—in order to be able to watch his trigger finger and his eye. In the later stages of instruction firing, the coach may be withdrawn from the firing line to observe his pupil from a point in rear; this affords him an opportunity to observe the pupil's performance while the latter is working alone, as he will be in record

practice. The pupil's errors should be noted and brought to his attention at the completion of the score.

(3) *Watching the eye.*—Errors in trigger squeeze, which are the most serious and the hardest to correct, can be detected by watching the pupil's eye. If his eye can be seen to close as the rifle goes off, it is because he knew when it was going off and consequently was not squeezing the trigger properly. The explosion and the shock cause a man to wink, but this wink cannot be seen, owing to the sudden movement of the head that takes place at the same time. If the firer can be seen to wink it is because he winked first and jerked the trigger afterward.

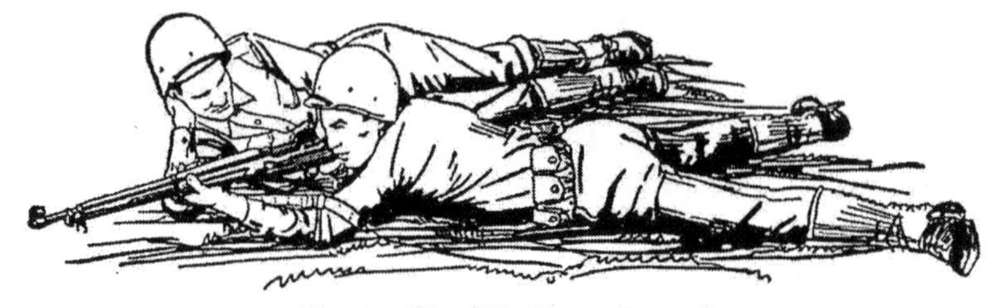

FIGURE 27.—Position of coach.

(4) *Preventing flinching.*—The best way to cure flinching is to prevent it. One of the most important duties of the coach is to detect indications that his pupil is about to flinch. An inclination to flinch can be detected by watching the pupil's head and eye while he is squeezing the trigger. If during the application of pressure to the trigger the pupil's eye begins to twitch and his head is gradually drawn away from the thumb or stock, it is a sure sign he will flinch. As soon as the coach notices these symptoms he should require the pupil to bring his rifle down from the firing position, rest a few seconds, and start all over again. While the pupil is resting the coach points out his errors, such as not taking up the slack, applying the squeeze with a series of impulses instead of continuous pressure, and concentrating on the discharge instead of on the sight picture.

(5) *Use of dummy cartridges in slow fire.*—If the pupil is seen to be flinching, or if he is doing poor or mediocre shooting, the coach first checks his aim by the aiming device. Having assured himself that the pupil is aiming correctly,

the coach has him turn his head aside while he, the coach, puts in a cartridge and shoves the bolt home. The coach frequently loads a dummy cartridge instead of a live round without letting the pupil know what he has done. Then the flinch, indicated by the shoulder being shoved forward at the same time that the trigger is pressed, will be evident even to the firer himself. The coach then proves to him by squeezing the trigger a few times, as explained in (6) below, that his poor shooting is caused by faulty trigger squeeze. He then proceeds to break the firer of flinching by a generous use of dummies.

(6) *Coach squeezing trigger* (fig. 28).—(a) To squeeze the trigger for the firer, the coach lies with his right elbow on the ground to steady his hand, places his thumb against the

FIGURE 28.—Coach squeezing trigger.

trigger, and his first finger against the back of the trigger guard. In this way he can apply pressure to the trigger by a pinching action of his thumb and first finger.

(*b*) The coach then watches the firer's back, and between 5 and 10 seconds after the firer begins to hold his breath he applies enough pressure to discharge the piece. Shots fired in this way are almost always accurately placed. After firing the piece a few times this way, the coach lets the firer try a few shots alone (the coach loading the rifle) to see if he can press the trigger the same way the coach pressed it, so as not to know just when the rifle will go off. Sometimes it is necessary to repeat this exercise, but most beginners can be permanently cured of the tendency to flinch by a few minutes of this kind of coaching. Old shots who are flinchers require more time and patience.

(7) *Duties of coach in slow fire.*—The coach observes the pupil carefully and corrects all errors. He pays particular attention to see—

(*a*) That the sights are blackened and set at the correct range.

(*b*) That the ammunition is free from dirt.

(*c*) That the pupil has the correct position, gun sling properly adjusted, body at the proper angle, elbows correctly placed, and the cheek resting firmly on the thumb or against the stock.

(*d*) That the magazine is loaded from a clip in the correct manner.

(*e*) That the slack is taken up promptly.

(*f*) That the trigger pressure is continued from that point with a smooth, steady, continuous increase.

(*g*) That the pupil is holding his breath properly (the coach watches his back occasionally).

(*h*) That the sights are correctly alined (the coach watches through the aiming device occasionally).

(*i*) Whether or not the pupil flinches (by watching his eye, his head, or the front sight for flips).

(*j*) That the pupil calls his shot each time he fires.

(*k*) That the pupil keeps his score book correctly.

(8) *Duties of coach in rapid fire.*—The coach observes the pupil carefully and corrects all errors. He pays particular attention to see—

(*a*) That the sights are blackened and that they are set at the proper range.

(*b*) That the gun sling is properly adjusted.

(*c*) That the pupil assumes the correct position.

(*d*) That he takes up the slack promptly.

(*e*) That he squeezes the trigger properly.

(*f*) Whether or not he flinches.

(*g*) That he works the bolt rapidly.

(*h*) That he breathes correctly while working the bolt.

(*i*) That while working the bolt the pupil keeps his eye on the target, the rifle to his shoulder, and his elbows in place.

(*j*) That he reloads deliberately and accurately.

(9) *Additional practice.*—These operations follow each other, and the coach can watch each in turn. Any lack of a smooth and rapid bolt operation indicates insufficient pre-

paratory training, and additional preparatory rapid-fire practice therefore is given.

f. Individual precautions.—(1) *Slow fire.*—The pupil should take the following precautions during every slow-fire score:

(*a*) Be sure that both the front and rear sights of the rifle are properly blackened.

(*b*) Be sure that the rear sight is properly set and the aiming point determined for the first shot.

(*c*) Take the score book, with preliminary data properly filled in, to the firing point.

(*d*) Fire the first shot very carefully and then, if necessary, change the sights or the aiming point, or both, to bring the second shot into the bull's-eye.

(*e*) Plot each shot in the score book.

(*f*) Before changing the sight setting, or the aiming point, note the setting on the sights and determine the amount of change required from a study of the shots plotted in the score book. Record the corrected sight setting and the aiming point.

(*g*) Do not change the sights or the aiming point unnecessarily. If a bad shot is made closely following several good shots it is almost certain to be the fault of the firer.

(2) *Rapid fire.*—In rapid-fire practice the pupil sees that his sights are blackened and properly set. Upon completion of a score he carefully plots each shot in the score book and promptly records any changes in sight setting or aiming point that may be necessary to center the group in the bull's-eye.

g. Use of instruments.—The use of binoculars, telescopes, sight-setting instruments, and instruments or devices for determining the force and direction of the wind is authorized and encouraged during instruction practice.

h. Procedure.—The procedure prescribed in paragraph 57 for record practice is applicable to instruction practice with the following exceptions:

(1) Scores are not required to be kept in the pits.

(2) Only such officers and noncommissioned officers are on duty in the pits as are necessary to preserve order and insure efficient pit service.

(3) The manner in which the scores are kept on the firing line is discretionary with the organization commander.

■ 57. RECORD PRACTICE.—*a. General.*—(1) The purpose of record practice is to test the soldier's skill as a rifleman and to determine his qualification. The qualification courses are prescribed in section III.

(2) The sequence in which the scores are fired in record practice is discretionary with the officer in charge of firing.

(3) Whenever practicable during record practice the officers required for duty in the pits are detailed from troops not firing.

b. Organization of firing line.—(1) *General.*—The firing line is organized to insure safe and orderly conduct and to facilitate supervision of the firing by the officer in charge of firing and his assistants. The distances specified in (2) below should be used as a guide only and may be modified at the discretion of the officer in charge of firing to meet local conditions.

(2) *Establishments.*—(*a*) Scorers are stationed in rear of the firing line and close to the soldier being scored.

(*b*) The ammunition line is 5 yards in rear of the firing line.

(*c*) Telephone operators are 5 yards in rear of the ammunition line.

(*d*) Soldiers awaiting their turn to fire (*ready line*) are 5 yards in rear of the line of telephone operators.

(*e*) Rifle rests and cleaning racks are 10 yards in rear of the ready line.

(3) *Assignments.*—Individuals who are to fire are assigned targets and the order in which they will take turn in firing the several scores, that is, first order, second order, etc.

c. Pit details.—The details for the supervision, operation, marking, and scoring of targets during record practice consist of—

(1) One commissioned officer assigned to each two targets. When it is impracticable to detail one officer to each two targets, an officer is assigned to supervise the marking and scoring of not to exceed four targets. In this event the pit scores are kept by the noncommissioned officer in charge of

each target, who signs the score card as soon as a complete score is recorded.

(2) One noncommissioned officer assigned to each target to direct and supervise the markers. This noncommissioned officer is selected from a company other than the one firing on the target which he supervises. If this is not possible the officer assigned to the target exercises special care to insure correct scoring.

(3) One or two privates assigned to operate and mark each target. These privates may be selected from the organization firing on the target to which they are assigned.

d. Score cards and scoring.—(1) Two score cards are kept, one at the firing point (W. D., A. G. O. Form No. 83), and one in the pit (W. D., A. G. O. Form No. 83-1).

(2) Entries on all score cards are made in ink or with indelible pencil. No alteration or correction is made on the card except by the organization commander, who initials each alteration or correction made.

(3) The scores at each firing point are kept by a noncommissioned officer of some organization other than that firing on the target to which he is assigned. If this is not possible, company officers exercise special care to insure correct scoring. As soon as a score is completed the score card is signed by the scorer, taken up and signed by the officer supervising the scoring, and turned over to the organization commander. Except when required for entering new scores on the range, score cards are retained in the personal possession of the organization commander.

(4) In the pit the officer keeps the scores for the targets to which he is assigned. As soon as a score is completed he signs the score card. He turns these cards over to the organization commander at the end of the day's firing. The organization commander checks the pit records against the firing line records. If the two records differ, the provisions of AR 345-1000 will be complied with.

(5) Upon completion of record firing and after the qualification order is issued, the pit score cards of each man are attached to his official score card kept at the firing point. For records and reports of qualification see AR 345-1000.

e. Marking.—(1) *Slow fire.*—(*a*) The value of the shot is indicated as follows:

1. A bull's-eye, with a white disk.

2. A four, with a red disk.

3. A three, with a black and white disk.

4. A two, with a black disk.

5. A miss or a ricochet hit, by waving a red flag across the front of the target.

(*b*) The exact location of the hit is indicated by placing in the shot hole a spotter of size appropriate to the distance from the firing point. The center of the marking disk is placed over the spotter in signaling hits. No spotters are required on 1,000-inch ranges.

(2) *Rapid fire on target D.*—(*a*) The same disks are used to indicate the value of hits as in slow fire.

(*b*) Spotters are placed in the shot holes before the target is run up for marking.

(*c*) The marking begins with the hits of highest value, the center of the disk being placed over the spotter, then swung off the target and back again to the next spotter, care being taken each time to show only the face of the disk indicating the value of the shot being marked. The marking should be slow enough to avoid confusing the scorer at the firing point. When a spotter covers more than one shot hole the disk is placed over it the required number of times.

(*d*) Misses and ricochet hits are indicated by slowly waving the red flag across the face of the target one time for each miss or ricochet hit.

f. Procedure.—(1) *Slow fire.*—(*a*) *On firing line.*

1. Only one person is assigned to a target in each order.

2. As the value of each shot is signaled, the scorer announces in a tone loud enough to be heard by the firer the name of the firer, the number of the shot, and the value of the hit. He then records the value of the hit on the score card of the individual who is firing.

3. When someone fires on the wrong target and the target is marked before the individual assigned to that target has fired, the scorer notifies the officer in charge, who notifies the officer assigned to that target in the pit to disregard the shot. This precaution is necessary to prevent errors in the pit records.

4. When an individual fires on the wrong target he is not scored a miss until the target to which he is assigned has been pulled down and the miss signaled from the pit.

5. If the target is not half masked at the completion of a score on that target, or if it is half masked at the wrong time, the officer in charge of that firing point adjusts the matter at once over the telephone. This precaution is necessary to prevent the error from being carried on through the scores that follow.

(*b*) *In the pit.*

1. The target is withdrawn and marked after each shot, except that on 1,000-inch ranges the targets are marked and removed after each 5 or 10 shots and replaced with new targets.

2. When a shot is fired at a target it is pulled down. The noncommissioned officer makes a pencil mark across the shot hole and indicates the location of the hit to the officer. The officer announces its value and records it on the score card. A spotter is then placed in the shot hole. The previous shot hole, if any, is pasted, and the target is run up and marked. The noncommissioned officer supervises the marking of each shot. The officer also exercises general supervision over the marking.

3. When the pit score card indicates that a score has been completed, the target is half masked for about 30 seconds as a signal of completion to the firing line. At the end of the 30 seconds the target is pulled fully down, the spotter removed, the shot hole pasted, and the target run up for the beginning of a new score.

4. When a target frame is used as a counterweight for a double sliding target, the blank side of the frame will be toward the firing line.

(2) *Rapid fire on target.*—(*a*) *On firing line.*

1. Only one person is assigned to a target in each order. The loop or hasty sling (as required) may be adjusted on the arm prior to the start of the exercise.
2. When all is ready in the pit a red flag is displayed at the center target. At that signal the officer in charge of the firing line commands: LOAD. The rifles are loaded and locked.
3. The officer in charge of the firing line then calls so that all may hear, "Ready on the right?" "Ready on the left?" Anyone who is not ready calls out, "Not ready on No. ——."
4. All being ready on the firing line, the officer in charge commands: READY ON THE FIRING LINE. Rifles are unlocked and the position of READY is assumed. The telephone orderly notifies the pit, "Ready on the firing line."
5. The flag at the center target is waved and then withdrawn. Five seconds after the flag is withdrawn the targets appear, remain fully exposed for the prescribed period of time, and are then withdrawn. The firer takes the prescribed position as soon as the targets appear and fires or attempts to fire 10 shots, reloading from a full clip taken from the belt. If any individual fails to fire at all he is given another opportunity.
6. As soon as the targets are withdrawn the officer in charge commands: UNLOAD. All unfired cartridges are removed from the rifle and the bolts are left open. The men remain in position on the firing line until they are ordered off by the officer in charge.
7. As each shot is signaled it is announced as follows: "Target No. —— 1 five, 2 fives, 3 fives, 1 four, 2 fours, 3 fours, 4 fours, 1 three, 1 miss, 2 misses."

The scorer notes these values on a pad and watches the target as he calls the shot. After the marking is finished he counts the number of shots marked, and if the number is more or less than 10, he calls "Re-mark No. ——." If 10 shots have been marked he then enters the score on the soldier's score card and totals it as follows: 5 5 5 4 4 4 4 3 0 0 equals 34.

(b) *In the pit.*

1. The time is regulated in the pit by the officer in charge.
2. When all is ready in the pit the targets are fully withdrawn and a red flag is displayed at the center target.
3. When the message is received that the firing line is ready, the red flag at the center target is waved and withdrawn and the command READY is given to the pit details.
4. Five seconds after the red flag is withdrawn the targets, by command or signal, are run up, left fully exposed for the prescribed period of time, and then withdrawn.
5. The officers in the pit examine each of their targets in turn, announce the score, and record it on the pit score card. Spotters are then placed in the shot holes and the targets run up and marked. The noncommissioned officer supervises the marking of each shot. The officer exercises general supervision over the marking.
6. The targets are left up for about 1 minute after being marked and are then withdrawn, pasted, and made ready for another score. They may be left up until ordered pasted by the officer in charge of the firing line.
7. If more than 10 hits are found on any target, the target is not marked unless all of the hits have the same value. The officer in charge of the firing line is notified of the fact by telephone.

(3) *Rapid fire on target D (rifle) 1,000-inch range.*—(*a*) *1,000-inch range with target pit.*—Rapid fire from the

standing position to prone, sitting, and kneeling is conducted in the same manner as prescribed for target D, except that the miniature targets are removed and replaced with new targets after marking.

(*b*) *1,000-inch range without target pit.*—The following provisions govern:

1. If the targets are covered by a curtain which can be opened to expose the face of the target and closed again to conceal it, or if the targets operate on a pivot, the rapid fire is conducted as closely as practicable in conformity with the method set forth above for a 1,000-inch range with target pit.

2. If the targets are exposed all the time, rapid fire from standing to prone, kneeling, and sitting is conducted by the officer in charge, who commands: 1. LIE DOWN (KNEEL OR SIT DOWN), 2. COMMENCE FIRING, 3. CEASE FIRING. Time is taken from the first command.

3. After the command UNLOAD and after all unfired cartridges are removed from the rifles and the bolts are open, the officer in charge directs the target detail to mark, remove, and replace the targets.

g. Use of telephones.—(1) Telephones are used for official communication only.

(2) No one will ask over the telephone for information as to the name or organization of any person firing on any particular target, and no information of this nature will be transmitted.

(3) The following expressions are used over the telephone in the situations indicated:

(*a*) When a shot has been fired and the target has not been withdrawn from the firing position, "Mark No. ——."

(*b*) When a shot has been fired and the target withdrawn from the firing position but not marked, "Disk No. ——."

(*c*) When the target has been withdrawn from the firing position and marked, but the value of the shot has not been understood, "Re-disk No. ——."

(*d*) When the firing line is ready for rapid fire, "Ready on the firing line."

(*e*) When a shot is marked on a target and the person assigned thereto has not fired, "Disregard the last shot on No. ——."

h. Miscellaneous rules governing record practice.—(1) *Identity of firer to be unknown to personnel in pit.*—Officers and men in the pit should not know who is firing on any particular target, and will not attempt to obtain this information; other officers and men will not transmit such information to personnel in the pit.

(2) *Coaching prohibited.*—Coaching of any nature, after the firer takes his place on the firing point, is prohibited. No person will render or attempt to render the firer any assistance whatever while he is taking his position or after he has taken his position at the firing point. Each firer must observe the location of his own hits as indicated by the marking disk or spotters. *The above prohibitions on coaching may be suspended during periods of mobilization.*

(3) *Use of instruments.*—(*a*) The use of binoculars, telescopes, and sight-setting instruments is authorized and encouraged.

(*b*) The use of instruments or devices for determining the force and direction of the wind is prohibited during record practice.

(4) *Shelter for firer.*—Sheds or shelter for the firer are not permitted at any range.

(5) *Restrictions as to rifle.*—Troops use the rifle with which they are armed. The rifle is used as issued by the Ordnance Department. The use of additional appliances, such as temporary shades for the sights, spirit levels, and orthoptic eyepieces is prohibited. The sights may be blackened. Any authorized size of peep sight issued by the Ordnance Department may be used. Small arms and appliances issued by the Ordnance Department for test and report will not be used for determining classification.

(6) *Trigger pull.*—The trigger pull must be at least 3 pounds and before record firing the pull must be tested (with the barrel vertical) by an officer.

(7) *Ammunition.*—The ammunition used is the service cartridge as issued by the Ordnance Department, unless the use of other ammunition is authorized.

(8) *Cleaning.*—Cleaning is permitted only between scores.

(9) *Use of gun sling.*—The gun sling will be used in connection with *one arm only.* For the purpose of adjustment for shooting, neither end will be removed from either sling swivel. No knot will be tied in the sling. The sling itself will neither be added to nor modified in any way. When the loop sling is authorized it may be adjusted (secured) to the arm prior to the start of the exercises.

(10) *Pads and gloves.*—(*a*) To reduce the shock of recoil, to prevent bruising the elbows, and to prevent irritation of the upper arm by the gun sling, pads of moderate size and thickness may be worn on either shoulder, on both elbows, and on either upper arm. Pads of such size, thickness, or construction as to form artificial support for the rifle are prohibited. Shoulder pads so designed by means of excessive size or thickness, quilting, rolls, ridges, or other devices as to aid materially in retaining the rifle butt in the firing position against the shoulder are prohibited. The use of a hook, small roll, or ridge on the sleeve of the shooting coat or shirt to keep the sling in place on the arm is prohibited.

(*b*) A glove may be worn on either hand provided it is not used to form an artificial support for the rifle.

(11) *Loading pieces.*—Pieces are not loaded except by command or until the position for firing has been taken.

(12) *Warming or fouling shots.*—No warming or fouling shots are allowed.

(13) *Action in case of disabled rifle.*—Should a breakage occur the rifle is repaired or a different rifle is substituted. If a different rifle is substituted, the firer is allowed to zero the substituted rifle and then refire the exercise.

(14) *Shots cutting edge of bull's-eye or line.*—Any shot cutting the edge of the figure or bull's-eye is signaled and recorded as a hit in the figure or the bull's-eye. Because the limiting line of each division of the target is the outer edge of the line separating it from the exterior division, a shot touching this line is signaled and recorded as a hit in the higher division.

(15) *Slow-fire score interrupted.*—If a slow-fire score is interrupted through no fault of the person firing, the unfired shots necessary to complete the score are fired at the first opportunity thereafter (see (13) above).

(16) *Misses.*—In all firing, before any miss is signaled, the target is withdrawn from the firing position and carefully examined by an officer, if an officer is on duty in the pit. Whenever the target is run up and a miss is signaled, it is presumed that this examination has been thoroughly made. No challenge of the value signaled is entertained and no resignaling of the shot is allowed.

(17) *Accidental discharges.*—All shots fired by the soldier after he has taken his place at the firing point (it being his turn to fire and the target being ready) are considered in his score even if his piece was not directed toward the target or was accidentally discharged.

(18) *Firing on wrong target.*—Shots fired upon the wrong target are entered as a miss upon the score of the man firing, no matter what the value of the hit upon the wrong target may be. In rapid fire the soldier at fault is credited with only such hits as he makes on his own target.

(19) *Two shots on same target.*—In slow fire, if two shots strike a target at the same time or nearly the same time, both are signaled; if one of these shots was fired from the firing point assigned to that target, the hit having the highest of the two values signaled is entered on the soldier's score and no record is made of the other hit.

(20) *Withdrawing target prematurely.*—In slow fire, if the target is withdrawn from the firing position just as the shot is fired, the scorer at that firing point should at once report the fact to the officer in charge of the scoring on that target. That officer investigates to see whether the occurrence is as represented. If he is satisfied that the target was moved as claimed, he will direct that the shot be not considered and that the man fire another shot.

(21) *Misfires in rapid fire.*—In the event of a misfire during rapid fire, the soldier ceases firing immediately, the target is not marked, and the score is repeated.

(22) *Unfired cartridges in rapid fire.*—In rapid fire, each unfired cartridge is recorded as a miss. If the number of

hits marked exceeds the number of rounds fired, the soldier firing on that target is credited with the hits of highest value corresponding to the number of rounds fired.

(23) *Disabled rifle in rapid fire.*—If, during the firing of a rapid-fire score, the rifle becomes disabled through no fault of the soldier, the pit officer is directed to disregard the score, the target is not marked, and the firer refires the score. The breaking of a clip in reloading does not entitle the soldier to another score.

(24) *More than 10 hits in rapid fire.*—When a target has more than 10 hits in rapid fire, the target is not marked, and the soldier firing on that target refires his score; except, however, that when all the hits on target D or on target D (rifle) 1,000-inch range have the same value, the target is marked and the value of the hits is credited to the soldier for each shot he has fired.

Section V

EQUIPMENT; KNOWN-DISTANCE TARGETS AND RANGES; RANGE PRECAUTIONS

■ 58. Equipment.—*a. Equipment for preparatory marksmanship training.*—(1) *General.*—The use of complicated apparatus which cannot be readily improvised from materials at hand is prohibited during preparatory marksmanship training. The simple apparatus described below is ample for all purposes.

(2) *Equipment for each four men.*

1 sighting bar, complete.
1 (each) large peep sight, front sight, and bull's-eye, made of cardboard.
1 rifle rest.
1 small sighting disk.
2 small aiming targets (rifle targets A and D, 1,000-inch range are suitable).
1 10-inch sighting disk.
1 small box, approximately the size of an ammunition box.
1 frame covered with blank paper for long-range triangles.

2 sandbags.
1 pencil.
20 rounds corrugated dummy ammunition.
4 score books (one per man).
1 form showing state of training.
4 empty cartridge clips.
Material for blackening sights.

(3) *Equipment for general use.*
1 rapid-fire target, with curtain, for each three squads.
1 each A, B, and D targets, on frames, for scorebook exercises.
1 aiming device for each squad.
Cleaning and preserving materials.

(4) *Preparation of sighting bar.*—(*a*) Provide a bar of wood about 1 by 2 inches and 4½ feet long. Cut two thin slots 1 inch deep across the edge. Place one slot 5½ inches from the end and the other 26 inches from the same end of the bar (fig. 29② and ③).

(*b*) Make a front sight ½ inch wide, of thin metal 1½ by 3 inches, with wing guards ⅛ inch wide extending ¼ inch on both sides of front sight and ¼ inch above height of the front sight. Bend in the shape of an L and tack it to the edge of the bar between the two slots and ½ inch from the slot nearest the end (fig. 29①). Have the leg of the L project above the bar ½ to ¾ of an inch (fig. 29④(A) and (B)).

(*c*) Make an eyepiece from a piece of tin or zinc 3 by 7 inches (fig. 29④(C), (D) and (E)). Cut along the dotted lines to form a shape shown in figure. Tack this eyepiece to the end of the bar farthest from the slots so that the top of the eyepiece extends 1 inch above the top of the bar (fig. 29 ③). Make a round hole 0.03 inch in diameter in the middle of the eyepiece ½ inch above the bar.

(*d*) Make a peep sight of thin metal or cardboard, measuring 3 by 3 inches, and cut a round hole ¾ inch in diameter in its center (fig. 29④(F)).

(*e*) Cut a piece of thin metal or cardboard, measuring 3 by 3 inches and painted white, and paste or paint a black bull's-eye ½ inch in diameter on the center (fig. 29④(G)).

(*f*) Place two pieces of tin 1 inch wide and 3 inches long

in each slot. Fold the loose ends away from each other and tack them to the sides of the bar (fig. 29③).

(*g*) Blacken the eyepiece, the front sight, the rear sight and the top of the bar.

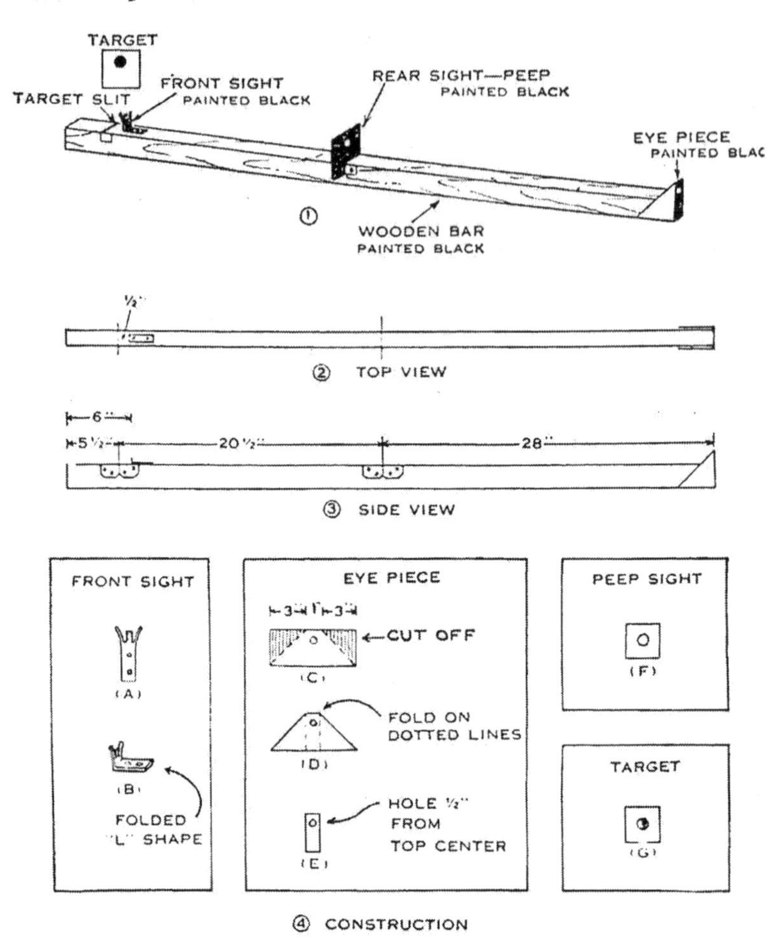

FIGURE 29.—Construction of sighting bar.

Wooden bar—1 by 2 inches by 4 feet 6 inches (approximate).
Eyepiece—Thin metal, 3 by 7 inches; hole, 0.03-inch diameter.
Rear sight—Thin metal or cardboard, 3 by 3 inches; hole in center, ¾-inch diameter.
Front sight—Thin metal, 1 by 3 inches, bent L shape.
Target—Thin metal or cardboard, 3 by 3 inches, painted white. Black bull's eye ¾-diameter in center.
Slits—1 inch deep, may be lined with thin metal strips.

(5) *Preparation of rifle rest.*—An empty ammunition box or any other well-made box of suitable size, with notches cut in the ends to fit the rifle closely, makes a good rifle rest. The rifle is placed in these notches with the trigger guard close to and outside of one end. The sling is loosened and pulled to one side. The box may be half filled with earth or sand to give it more stability.

(6) *Preparation of sighting disks.*—Sighting disks are of three sizes. The disk to be used at a distance of 50 feet is about 3 inches in diameter. The disk is made of tin or cardboard and mounted on a handle as shown in figure 30. The bull's-eye is mounted on a background of clean, white paper. The disks to be used at 200 and 500 yards are, respectively, 10 and 20 inches in diameter. These disks are painted black and mounted on white handles which are 4 or 5 feet long. All bull's-eyes are black and circular and have a hole in the center large enough to admit the point of a pencil.

b. Range equipment.—(1) *Used at firing point.*

Cleaning racks.
Scorers' tables.
Binoculars (1 per target).
Score cards.
Score board.
Cleaning and preserving materials.
Material for blackening sights.
Score books.
Indelible pencils.
Containers for empty cartridge cases.
Telephones.

(2) *Used in pit.*

Pit record cards.
Indelible pencils.
Telephones.
Ten 3-inch spotters per target.
One 6-inch spotter per target.
One red flag per target.
Marking disks.
Pasters.
Paste.

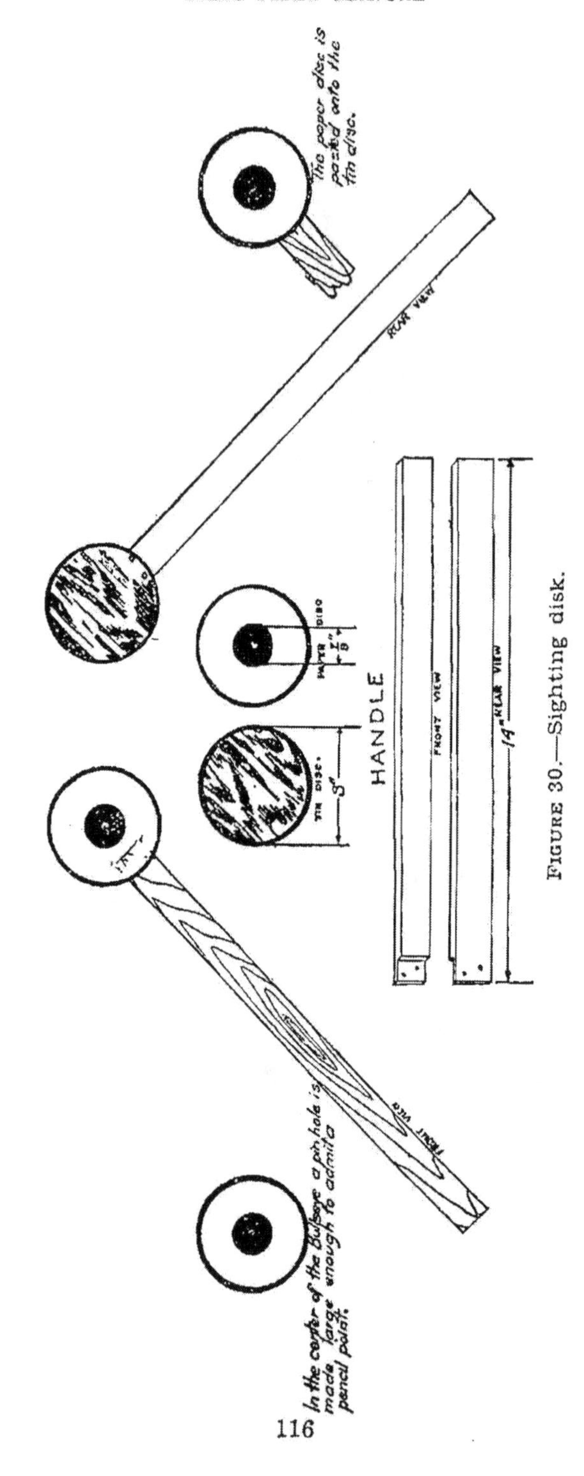

FIGURE 30.—Sighting disk.

■ 59. TARGETS.—The specifications for marksmanship targets, together with the value of hits in their divisions, are as follows:

a. Target A, the short-range target, used for 200 and 300 yards, is a rectangle 6 feet high and 4 feet wide; black circular bull's-eye, 10 inches in diameter, value of hit, 5; center ring, 26 inches in diameter, value of hit, 4; inner ring, 46 inches in diameter, value of hit, 3; outer, remainder of target, value of hit, 2.

b. Target B, the midrange target, used for 500 yards, is a square 6 feet on a side; black circular bull's-eye, 20 inches in diameter; center ring, 37 inches in diameter; inner ring, 53 inches in diameter; outer, remainder of target. Value of hits same as on target A.

c. Target D, the rapid-fire target, is a square 6 feet on a side and has in its middle a black silhouette representing a soldier in the prone position. Value of hits in the figure, 5; in the space immediately outside the figure, 4; in the space immediately outside the 4 space, 3; remainder of the target, 2.

d. Target A (rifle) 1,000-inch range and target D (rifle) 1,000-inch range are a reduction of targets A and D, respectively, from 200 yards to 1,000 inches. Values of the hits on the reduced targets are the same as for targets A and D.

■ 60. KNOWN-DISTANCE TARGET RANGES.—*a. General.*—There are two classes of ranges: class A ranges, which are more or less limited in extent and are equipped for known-distance practice; and class B ranges, which are of extended area and diversified terrain and are used for combat firing. The following subparagraphs refer to class A ranges only.

b. Rules for selection.—Since the nature and extent of the ground available for target practice as well as the general climatic conditions may vary widely for different military posts, it is not possible to prescribe hard and fast rules governing the selection of ranges. The best that can be done is to indicate certain broad, general matters which must be considered. In view of the range and penetration of the bullet of the U. S. rifle, M1917, it will be found necessary in many posts to conduct target practice several miles away from the post proper. This requires the establishment of a camp at or near the range.

c. Security necessity.—For posts of limited extent situated in thickly settled localities, the first condition to be fulfilled is security for those who live or work near the range. This requirement can be met for class A ranges by selecting ground where a natural butt is available or by making an artificial butt sufficiently extensive to stop wild shots. See AR 750–10 for information concerning danger areas.

d. Direction of range.—If possible, a range should be so located that the firing is toward or slightly to the east of north. This orientation gives a good light on the face of the targets during the greater part of the day. However, security and suitable ground are more important than direction.

e. Best ground for class A range.—Smooth, level ground or ground with a moderate slope is best adapted for a range. The targets should be on the same level with the firer or only slightly above him. Firing downhill should be avoided.

f. Size of range.—The size of the range is determined by its plan and by the number of troops that are to fire over it at a time. There are two general plans used in range construction: one with a single target pit and with firing points for each range; the other with firing points on one continuous line and with the target pits for the various ranges echeloned. The latter type requires more ground and is not so useful for training troops.

g. 1,000-inch range.—There are two classes of 1,000-inch ranges—those with a danger area behind the backstop and those without. Where possible, open 1,000-inch ranges for use in cities will be so constructed that no danger area exists behind the backstop. In addition, they should be so sited that sparsely settled territory is behind the backstop and so located that the range will not be a noise nuisance. 1,000-inch ranges requiring a danger area behind the backstop must meet the same security requirements as class A ranges.

h. Principles governing construction.—(1) *Intervals between targets.*—To reduce to a minimum the amount of labor required in preparing the range, the targets should be no farther apart than necessary to reduce the probability of firing on the wrong target. As a general rule, the intervals between targets are equal to the width of the targets themselves; that is, at short and midrange, 6 feet; at long range, 12 feet. Where as many targets as possible must be crowded into a limited space, this interval may be halved without materially affecting the value of the instruction.

(2) *Protection for markers.*—(*a*) On all ranges, protection is provided for the pit details by excavating a pit for the targets, or by constructing a parapet in front of them, or by a combination of pit and parapet.

(*b*) Where there are several targets in a row, the shelter should be continuous. It must be high enough to protect the markers. The parapet may be of earth, with a timber or concrete revetment of sufficient thickness to stop bullets, and rise from 7½ to 8 feet above the ground or platform on which the markers stand.

(3) *Artificial butts.*—If an artificial butt is constructed as a bullet stop, it should be of earth not less than 30 feet high and with a slope of not less than 45°. It should be extended at least 5 yards beyond the outside targets and should be placed as close behind the targets as possible. The slopes should be sodded.

(4) *Hills as butts.*—A natural hill to form an effective butt should have a slope of not less than 45°; if it is naturally more gradual, it should be cut into steps, the face of each step having a slope of not less than 45°. As a temporary expedient, the face of the hill may be plowed perpendicularly to the range, but as the bullets soon cut down the furrows this must be frequently repeated to prevent ricochets.

(5) *Numbering of targets.*—Each target should be designated by a number. The numbers for ranges up to 600 yards should measure at least 6 feet in height and should be painted black on a white background. Arabic numerals 6 feet tall can always be quickly recognized. They should be placed on the butt behind each target or on the parapet in front, and close enough (above or below) so the firer can see the number while aiming at the target.

(6) *Measuring the range.*—The range should be carefully measured and marked with stakes at the firing points in front of each target. These stakes should be about 12 inches

above the ground and painted white. They should have in black figures the number of the corresponding target and its distance. Particular care should be taken to place each stake so it is parallel to the face of its own target.

(7) *Ranges parallel.*—The different ranges for the same distance should all be parallel, so that similar conditions with respect to wind and light may exist. It is not essential, however, that the ranges used for long-distance shooting parallel those used for ordinary marksmanship practice.

(8) *Firing mounds.*—If it becomes necessary to raise a firing point on account of low ground, a low mound of earth should be made. The mound should be level, sodded, and not less than 12 feet square. If the entire firing line is raised, the firing mound should be level, sodded, and not less than 12 feet wide on top.

(9) *Pit shed.*—A small house or shed should be built in or near the target pit, in which to store the marking disks and signal flags and spare parts for making immediate repairs on the target frames. It should be large enough to afford a shelter for the markers in case of a sudden storm.

(10) *Danger signals.*—A socket for the staff of the danger signals should be placed on the markers' shelter in front of each target and so inclined that the flag will always fall clear of the staff and be readily seen. This flag is always displayed when the target is in place and not in use. In addition to the danger signals at the targets, a scarlet streamer is displayed from a prominent point on all ranges and at all times during firing to warn passers-by when firing is in progress. These signals are not to be placed in such a position as to serve as streamers for judging wind on the range. They should be placed on the roads or on the crest of the hill, where they can be seen plainly by those passing.

(11) *Range house.*—On large ranges where competitive firing is held, a house containing a storeroom and several office rooms should be erected in some central place near the range. Facilities to enable visitors to witness the firing should also be provided.

(12) *Telephone service.*—Ranges should be equipped with a telephone system connecting the target pit with each firing point, the range house, and the post or camp proper. The

number of telephones should not be less than one to each 10 targets.

(13) *Electric bells.*—On large ranges it is desirable to install for each five targets an electric bell that can be controlled from a central point in the pit. Such a bell adds materially to the celerity and uniformity of target manipulation for rapid fire.

(14) *Covered ways between pits.*—If the pits are in echelon, covered ways or tunnels should be provided between the various pits. This construction allows the pit details to be shifted in safety without interrupting the firing.

(15) *1,000-inch range.*—An open 1,000-inch range requiring no danger area behind the backstop must meet the following minimum requirements:

(*a*) It must have a vertical bulletproof backstop and wing walls (natural or artificial) not less than 30 feet high. The wing walls must extend at least 15° on each flank. If artificial wing walls are used, they should be inclined toward the firing points at an angle of 15° with the backstop.

(*b*) In front of the firing points it must have a ricochet pit that provides at least a 4° slope downward from the normal line of fire from a prone position and extends to within 30 feet of the backstop and wing walls. If a vertical cliff or wall over 40 feet high is available as backstop, no ricochet pit need be provided.

■ 61. RANGE PRECAUTIONS.—See AR 750–10.

SECTION VI

SMALL-BORE PRACTICE

■ 62. OBJECT.—The object of small-bore practice is to provide a form of marksmanship training with the caliber .22 rifle and ammunition which represents the application of the principles taught in the preparatory exercises. Small-bore practice provides an excellent means of improving the shooting of organizations and sustaining interest in marksmanship throughout the year. The firing of this course enables the company commander to visualize the state of training of his command and to concentrate his efforts on the training of those who are most deficient.

■ 63. VALUE.—The chief value of small-bore practice lies in the fact that it is convenient, interest sustaining, and economical. It does not have the full value of caliber .30 practice because of the absence of recoil, but on account of its convenience and the saving in the cost of ammunition, it is a valuable part of marksmanship training.

■ 64. CONTINUOUS SMALL-BORE PRACTICE.—Small-bore practice may be carried on throughout the year, subject to limitations in ammunition allowances. Men who have not been instructed in the shooting methods prescribed in this manual must be given a thorough course of preparatory instruction before being permitted to fire on the small-bore range. Small-bore practice is organized and supervised in accordance with the methods of instruction prescribed in this manual.

■ 65. COURSES.—When ammunition allowances, time, and facilities permit, organizations may fire one of the small-bore courses outlined below.

a. Course E.—(1) *Instruction practice.*

(*a*) *Short range.*

TABLE I.—*Slow fire*
(To zero rifle)

Range (feet)	Time	Shots	Targets	Position	Sling
50	No limit	5	SB-A-2	Prone or prone with sandbag	Loop.
50	No limit	5	SB-A-2	Sitting	Loop.
50	No limit	5	SB-A-2	Kneeling	Loop.
50	No limit	5	SB-A-2	Standing	Hasty.

TABLE II.—*Slow fire*

Range (feet)	Time	Shots	Targets	Position	Sling
50	No limit	10	SB-A-2	Standing	Hasty.
50	No limit	10	SB-A-3	5 kneeling; 5 sitting	Loop.
50	No limit	10	SB-B-5	Prone	Loop.

TABLE III.—*Rapid fire*

Range (feet)	Time (seconds)	Shots	Targets	Position	Sling
50	60	10	SB-D-2	Standing to kneeling	Loop.
50	70	10	SB-D-3	Standing to prone	Loop.

NOTE.—When desired, tables I, II, and III may be fired at 1,000 inches by substituting target A, 1,000-inch, for targets SB-A-2 and SB-A-3; target B, 1,000-inch, for target SB-B-5; and target D, 1,000-inch, for targets SB-D-2 and SB-D-3.

(b) *Intermediate range.*

TABLE IV.—*Slow fire*

Range (yards)	Time	Shots	Targets	Position	Sling
50	No limit	10	SB-50 yards, A target.	Prone	Loop.
50	No limit	10	SB-50 yards, A target.	5 kneeling; 5 sitting	Loop.

TABLE V.—*Rapid fire*

Range (yards)	Time (seconds)	Shots	Targets	Position	Sling
50	70	10	SB-50 yards, D target.	Prone from standing	Loop.
50	70	10	SB-50 yards, D target.	Sitting from standing	Loop.
50	75	10	SB-50 yards, D target.	Kneeling from standing	Loop.

NOTE.—The firing included in tables IV and V is optional. If no 50-yard range is available, tables IV and V will be fired at 100 yards on the SB-100 yards, D target. This applies only to the E course.

(c) *Long range.*

TABLE VI.—*Slow fire*

Range (yards)	Time	Shots	Targets	Position	Sling
100	No limit	10	SB-100 yards, A target.	Prone	Loop.
100	No limit	10	SB-100 yards, A target.	5 kneeling; 5 setting	Loop.
100	No limit	10	SB-100 yards, D target.	Prone	Loop.

TABLE VII.—*Rapid fire*

Range (yards)	Time (seconds)	Shots	Targets	Position	Sling
100	80	10	SB-100 yards, D target.	Prone from standing	Loop.
100	80	10	SB-100 yards, D target.	Sitting from standing	Loop.

(2) *Record practice.*

TABLE VIII.—*Slow fire*

Range (yards)	Time	Shots	Targets	Position	Sling
100	No limit	10	SB-100 yards, A target.	5 kneeling; 5 sitting	Loop.
100	No limit	10	SB-100 yards, A target.	Prone	Loop.

TABLE IX.—*Rapid fire*

Range (yards)	Time (seconds)	Shots	Targets	Position	Sling
100	80	10	SB-100 yards, D target.	Prone from standing	Loop.
100	80	10	SB-100 yards, D target.	Sitting from standing	Loop.

b. Course F.—(1) *Instruction practice.* Fire tables I, II, III, of course E.

TABLE X.—*Rapid fire*

Range (feet)	Time (seconds)	Shots	Targets	Position	Sling
50	65	10	SB-D-3	Prone from standing	Loop.
50	70	10	SB-D-3	Sitting from standing	Loop.

(2) *Record practice.*

TABLE XI.—*Slow fire*

Range (feet)	Time	Shots	Targets	Position	Sling
50	No limit	10	SB-A-3	Kneeling	Loop.
50	No limit	10	SB-B-5	Prone	Loop.

TABLE XII.—*Rapid fire*

Range (feet)	Time (seconds)	Shots	Targets	Position	Sling
50	65	10	SB-D-3	Prone from standing	Loop.
50	70	10	SB-D-3	Sitting from standing	Loop.

CHAPTER 3

MARKSMANSHIP—MOVING GROUND TARGETS

SECTION I

GENERAL

■ 66. SCOPE OF TRAINING.—Rifle units are trained to fire at moving ground targets, such as lightly armored vehicles, trucks, and personnel at effective ranges.

■ 67. BASIC PRINCIPLES.—The principles presented in this chapter deal with firing at moving ground targets. In applying these principles the firer must adjust his aim and trigger squeeze to conform to the movement of the target.

a. Effective range.—Under ideal conditions, moving targets may be engaged at ranges beyond 600 yards, but effective results at such ranges are exceptional. Therefore, training in the technique of fire is normally limited to ranges of 600 yards or less.

b. Sights to be used.—Moving targets are seldom exposed for long periods and can be expected to move at maximum speed during periods of exposure. Accurate correction of sight setting is usually impracticable. Therefore, instruction in technique should favor the use of the battle sight. Corrections for range are made by adjustment of the aiming point on the target. The battle peep sight is habitually used against moving targets at close range.

c. Leads.—Targets which cross the line of sight at any angle are classified as crossing targets. In firing at crossing targets *the firer must aim ahead of the target* so that the paths of the target and bullet will meet. This distance ahead of the target is called the "lead." Targets which approach directly toward the firer or move directly away from him require no lead.

SECTION II

MOVING VEHICLES

■ 68. DETERMINATION AND APPLICATION OF LEADS.—*a.* The lead necessary to hit a moving vehicle depends upon the speed of the vehicle, the range to the vehicle, and the direction of movement with respect to the line of sight. For example, at 10 miles an hour a vehicle moves approximately its own length (5 yards) in 1 second. A rifle bullet moves about 400 yards in ½ second. Therefore to hit a vehicle moving at 10 miles an hour at a range of 400 yards, the lead should be 2½ yards, that is, one-half the target length or one-half lead.

b. Leads are applied by using as the unit of measure the length of the target as it appears to the firer. This eliminates the necessity for corrections due to the angle at which the target crosses the line of sight, because the more acute the angle the smaller the target appears and the less lateral speed it attains.

c. The following lead table is furnished as a guide:

TARGET LENGTHS

Miles per hour	400 yards or less	400–600 yards
10	½	1
20	1	2

■ 69. TECHNIQUE OF FIRE.—The following technique is suggested for firing at rapidly moving targets, using the battle peep sight:

a. Approaching or receding targets.—The firer holds his aim on the center of the target and squeezes off his shot.

b. Crossing targets.—(1) *At ranges less than 600 yards.*—The firer alines his sights on the bottom of the target at its rearmost point, and swings straight across it to the estimated lead. The rifle is kept swinging and the shot squeezed off while the proper lead is maintained.

(2) *At ranges of 600 or more yards.*—The firer proceeds as in (1) above except that he swings his point of aim across the top of the target.

c. Rapidity of fire.—Fire is delivered as rapidly as proper performance permits.

■ 70. PLACE IN TRAINING.—The technique of firing at moving vehicles should follow instruction in known-distance firing. When time and ammunition allowances permit, 1,000-inch and caliber .22 firing may be added as preliminary instruction.

SECTION III

MOVING PERSONNEL

■ 71. TECHNIQUE.—*a. Sight to be used.*—Under field conditions, moving personnel presents a fleeting target, more difficult to hit than a moving vehicle. This fact causes the use of the adjustable peep sight desirable for greater accuracy. However, the use of the battle peep sight is necessary when targets appear suddenly and there is no time for sight adjustment. Therefore the rifleman should be trained in the use of both sights for this type of firing.

b. Method of aiming.—An elaborate system of calculating leads is neither necessary nor desirable. The following general rule forms the basis for estimating the proper leads. For firing at a man walking across or at right angles to the line of fire, the points of aim at the various ranges are—

(1) At 100 yards, *aim at forward half of body.*

(2) At 200 yards, *aim at forward edge of body.*

(3) At 300 yards, *lead him one-half the width of his body.*

(4) At 400 yards, *lead him the width of his body.*

■ 72. PLACE IN TRAINING.—Instruction in this type of firing should follow instruction in known-distance firing and should immediately precede the training of the squad in technique of fire (musketry) when practicable. Proficiency in this type of firing depends largely upon the amount of training the individual receives in aiming, squeezing the trigger, and leading with appropriate speed.

SECTION IV

MOVING TARGETS, RANGES, AND SAFETY PRECAUTIONS

■ 73. GENERAL.—Instruction in firing at moving targets is an important phase of training. Units should be given this instruction whenever time and facilities permit.

a. Moving vehicles.—A suggested target for representing a moving vehicle is shown in figure 31. This target has the advantage of a low center of gravity, which prevents upsetting on rough ground and when making changes of direction. It is simple to construct. The method of operating it is shown in figures 32 and 33. Other types of targets may be improvised.

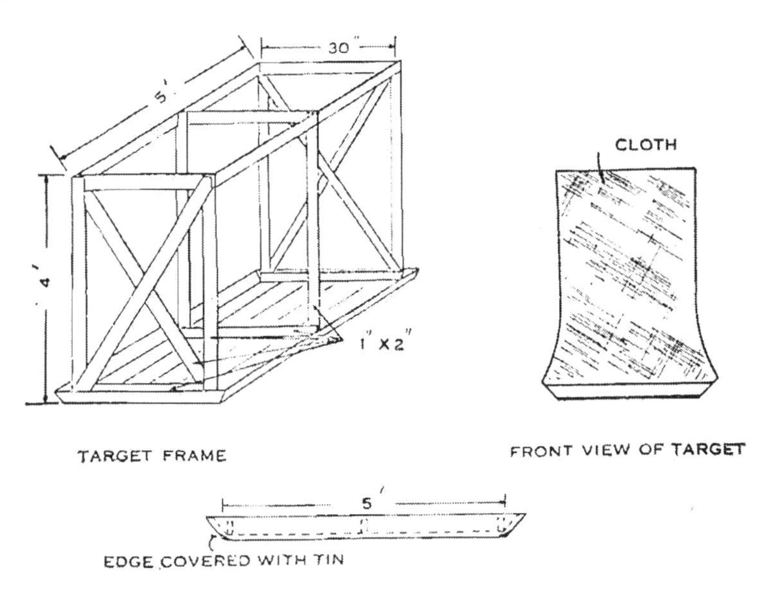

FIGURE 31.—Target frame for moving target range.

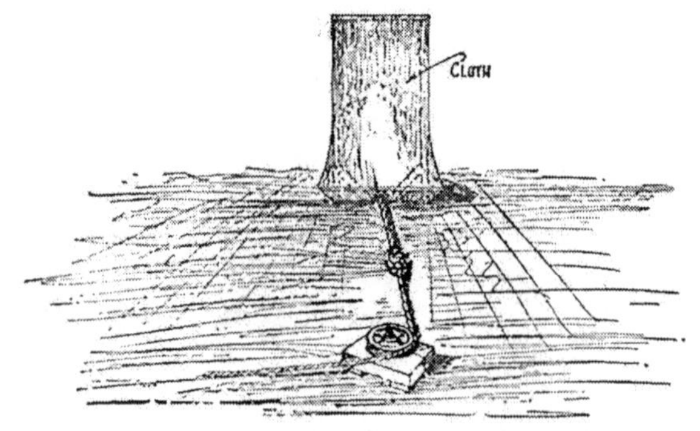

FIGURE 32.—Sled target covered with target cloth; pulley and trip knot for effecting changes of direction.

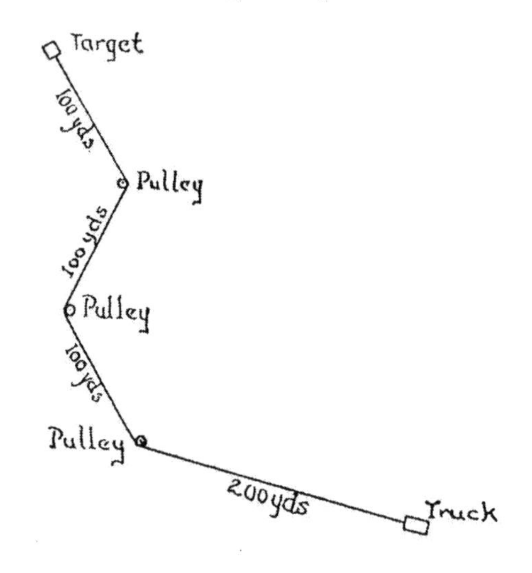

FIGURE 33.—Set-up for towing a target.

b. Moving personnel.—Any class A range is suitable for practice in firing at moving personnel. E targets on sticks carried by men in the pits are suitable.

■ 74. SAFETY PRECAUTIONS.—The following safety precautions listed must be rigidly adhered to:

a. Firing at moving targets is not permitted on any range until the safety angles have been carefully checked and markers have been placed so as to define clearly the right and left limits of fire.

b. Personnel of the towing trucks must operate far enough from the line of fire to be protected not only from direct hits but also from ricochets.

c. Trucks replacing targets on the course, or personnel making repairs, display red flags.

d. Target personnel do not leave the designated safety area until the signal or command to do so has been given.

e. The officer in charge of firing is responsible for seeing that—

(1) Rifles are not pointed in any direction that will endanger the target detail.

(2) Rifles are not pointed in such a direction that the fire will extend outside the prescribed limits outlined on the range.

(3) Firing ceases immediately upon the command CEASE FIRING.

(4) Rifles are cleared and bolts opened before the target detail is permitted to move on the range.

(5) The range is cleared before the command LOAD is given.

(6) The general safety precautions listed in AR 750–10 are complied with.

CHAPTER 4

MARKSMANSHIP—AIR TARGETS

SECTION I

AIR TARGETS FOR RIFLE

■ 75. GENERAL.—Combat arms take the necessary measures for their own immediate protection against low-flying hostile aircraft. Therefore all infantry troops must be fully trained and imbued with the determination to protect themselves against hostile aerial attacks without reliance upon other arms.

■ 76. CLASSIFICATION.—From the point of view of the rifleman, air targets may be classified as—

a. Overhead—those which pass over or nearly over the rifleman; or *nonoverhead*—those which do not pass over or nearly over the rifleman. Either of these types may be flying at a constant altitude or may be losing or gaining altitude.

b. Direct diving—those which dive directly toward a rifleman; or *direct climbing*—those which climb directly away from a rifleman.

SECTION II

TECHNIQUE OF FIRE

■ 77. GENERAL.—Airplanes which are suitable targets for rifle fire present very fleeting targets. They must be engaged promptly by all available weapons. Riflemen must be taught a simple method of firing on hostile low-flying airplanes. This section deals entirely with actual fire on hostile airplanes. Details of antiaircraft marksmanship training are contained in sections III, IV, and V.

■ 78. LEADS.—*a. General.*—In order to hit a moving target, such as an airplane in flight, it is necessary to aim an appropriate distance ahead of it and on its projected path of flight so that the target and the bullet will meet. This distance ahead of the airplane is called "lead." A lead must be taken in all firing except when the target is at extremely close range (100) feet, when it is diving directly at the firer, or flying directly from him.

b. Determination of leads.—(1) The lead necessary to engage a moving target depends upon—

(*a*) The speed of the target.

(*b*) The range to the target.

(*c*) The time of flight of the bullet.

(*d*) The direction of movement of the target with respect to the line of fire.

(2) When a target appears, it is impossible for riflemen or leaders of rifle units to consider all of the factors in (1) above and from these compute the lead required. Therefore, leads are computed and placed in lead tables for the use of leaders in training their units. (See par. 170*c*.)

c. Application of leads.—Although leads are originally computed in feet or yards, they are given in lead tables as target lengths. It is very difficult to estimate, with any degree of accuracy, a lead such as 40 or 50 yards at ranges from 600 to 100 yards. Therefore, the length of the target as it appears to the firer is used as the unit of measure for applying leads. The rifleman is trained to apply the length of the target, as it appears to him, along the projected path of the target to determine the aiming point for each shot. The number of times he applies this unit of measure is announced in a fire order or will be determined by methods explained in paragraph 80.

■ 79. TARGET DESIGNATION.—*a.* Aerial target designation should be given as routine training in training areas long before the area of probable hostile air attack is reached. Aerial targets for a single unit usually are clearly visible and few in number.

b. Attacking aviation will usually fly in a V-shaped formation of three airplanes or will operate individually. Should all the fire of a rifle unit be directed at one airplane, the normal dispersion will result in effective fire on the remaining airplanes of the formation. Therefore the normal method of target designation is to assign each of the three airplanes to an element of the rifle platoon during the training period. For example, the center squad may be assigned the leading airplane, the second squad the right airplane, and the third squad the left airplane. If less than three airplanes attack, the units not having a target assigned to them fire on the leading airplane. The assignment is not changed except in unusual circumstances.

c. The usual assignment of a target extends from its initial appearance until it passes beyond range. If a unit is attacked by a succession of groups of hostile airplanes, the leader causes his unit to cease fire at one group in time to bring fire on the following groups as they come within effective range.

■ 80. FIRE DISTRIBUTION.—*a*. The fire of rifle units must be distributed along the path of flight of the target as long as the target is within effective range. This is done as follows:

(1) For all targets, except direct diving or direct climbing, *aim and fire each shot with ten target-length leads.*

(2) For all direct diving or direct climbing targets *aim and fire each shot at the target.*

b. This method of fire distribution is based upon the fact that as the target is approaching or receding, the range and the leads are constantly changing. The lead used is the average of all leads necessary to engage a target between the extreme effective range of 600 yards and a minimum range of 100 yards.

c. The target considered in determining the lead of ten target lengths is a 30-foot airplane. In using this method for towed-target firing the lead should be changed in accordance with the length of the sleeve target.

d. It is impracticable for men to estimate airplane speeds with any degree of accuracy; therefore the speed of present day attack airplanes, which is approximately 300 miles per hour, is used. For speeds considerably greater or less than 300 miles per hour the lead should be changed proportionately. Experience has shown that this method of distribution gives results equal to or better than more accurate and more complicated methods.

■ 81. DELIVERY OF FIRE.—*a. Range.*—(1) The maximum effective range of rifle fire at air targets is approximately 600 yards. However, riflemen should take the firing position as soon as possible after receiving the warning of the approach of hostile airplanes and track the target until it comes within range.

(2) Training in estimating ranges of air targets is conducted by having individuals observe airplanes flying at known ranges. The individual bases his estimate on the appearance of the airplane at key ranges. In general, the following parts of an airplane are visible at the ranges indicated:

	Range (yards)
General outline	1,000
Wheels, rudder, wing struts, tail skid	700
Antenna and small projections from fuselage	500
Symbols, numbers, letters	200

b. Rate.—The rate of fire at aerial targets is about the same as the rapid-fire rate at ground targets. Everything must be done to increase the rate of fire without affecting its accuracy. Repeated tests have proved that rifle fire delivered faster than is consistent with proper aim and trigger squeeze results in waste of ammunition. Each shot must be aimed and squeezed. A well-trained rifleman can fire one shot in 3 or 4 seconds.

c. Sights used.—The use of the battle peep sight or the adjustable peep sight is impractical in firing at aerial targets because the rear sight wings obscure the target when aiming. Therefore, when engaging aerial targets the sight leaf is turned down, and the firer sights over the top of the ring of the battle peep sight in alining the front sight on the estimated lead.

d. Accuracy.—Combat experience indicates that the antiaircraft fire of trained riflemen is effective and should cause substantial losses to hostile air units.

e. Effect of caliber .30 fire on airplane.—(1) Various degrees of damage may be inflicted on an airplane by rifle fire. Hits on the cylinder walls and other important working parts are likely to stop an engine immediately. A hit through the metal propeller is also serious, since it throws the engine out of balance. Unless the bombs carried by the airplane are bulletproof, hits by armor-piercing small-arms bullets will detonate them. The pilot is especially vulnerable.

(2) There are many lesser ways in which infantry fire can damage an airplane. Holes through the crank case may cause the oil to drain out and the engine to freeze before the airplane returns to friendly territory. Hits of any kind require varying degrees of repair, even if they do not cause the destruction of the airplane.

Section III

ANTIAIRCRAFT MARKSMANSHIP

■ 82. General.—*a. Object of instruction.*—The object of antiaircraft marksmanship instruction is to train the rifleman in the technique of firing at rapidly moving aerial targets.

b. Basis of instruction.—(1) Prior to instruction in antiaircraft marksmanship the soldier should have completed a course of training in ground firing and have acquired the fundamentals of good shooting. To become a good antiaircraft marksman he must be able to apply the fundamentals of target practice in firing at rapidly moving targets and to perform the following operations with accuracy and precision:

(*a*) Apply the length of the target as a unit of measure in measuring the required lead.

(*b*) Aline the sights of the rifle on the required lead rapidly.

(*c*) Swing the rifle with a smooth, uniform motion so as to maintain the aim on the required lead while getting off the shot.

(*d*) Apply continuous trigger squeeze so as to fire in a minimum of time and without disturbing the aim.

(2) The correct performance of these four operations combined into one continuous, smooth motion when firing in any direction at rapidly moving aerial targets is the basis for the course of training outlined in this manual.

c. Sequence of training.—Antiaircraft rifle marksmanship is divided into preparatory exercises, miniature range practice, and towed-target firing.

d. Personnel to receive training.—All personnel of units whose primary weapon is the rifle will be trained in antiaircraft marksmanship to the degree permitted by available time and ammunition.

■ 83. PREPARATORY EXERCISES.—*a. General.*—(1) *Description.*—The preparatory exercises are designed to teach the soldier the fundamentals of antiaircraft rifle marksmanship and to drill him in these until the correct procedure becomes a fixed habit. The preparatory exercises consist of the following three steps which are completed prior to firing:

Position exercises.
Aiming and leading exercises.
Trigger-squeeze exercises.

(2) *Methods.*—A conference by the instructor precedes each exercise. He explains the exercise, the reason for it, and demonstrates it with well-trained and carefully rehearsed personnel. In order to awaken interest and stimulate enthusiasm, the preliminary instruction must be individual and thorough. Each man must understand and be able to explain each point.

(3) *Coaching.*—Whenever a man is in a firing position during the preparatory exercises and miniature range firing, he is assigned a coach whose duty is to watch him and point out errors. Soldiers are therefore grouped in pairs and alternate as coach and pupil. Unit leaders are the instructors; they supervise and prompt the coaches.

b. Organization.—With the targets and target ranges described in section VI, a group of 32 men per target is the most economical training unit. For the preparatory exercises this permits 16 men to work on each type target while the remaining 16 men act as coaches. Each group completes all preparatory training and instruction firing on its assigned target. Groups then change places. The preparatory training and instruction firing are then undertaken on a new type target. This procedure is followed until each man of each group has completed his instruction on each of the four types of targets.

■ 84. FIRST STEP: POSITION EXERCISES.—*a. General.*—The positions used in antiaircraft firing are those which can be assumed rapidly, which afford the maximum flexibility to the body for the manipulation of the rifle, and which do not require undue exposure of the firer. The kneeling and standing positions best meet these requirements (figs. 34 and 35); the kneeling position is the less vulnerable of the two.

FIGURE 34.—Rifleman in antiaircraft standing position.

b. Firing positions.—(1) Antiaircraft firing positions differ from those used in ground target firing in that—

(*a*) The sling is not used.

(*b*) The arms are not supported but move freely in any direction with the body.

(*c*) The hands grasp the piece firmly, the left hand near the lower band.

(*d*) The butt of the rifle is pressed firmly against the shoulder with the right hand, and the cheek is pressed against the stock.

FIGURE 35.—Rifleman in antiaircraft kneeling position.

(e) In the kneeling position the buttock does not rest on the heel, and the left foot is well advanced to the left front. The weight is slightly forward.

(2) The positions must be such that the rifle, the body from the waist up, the arms, and the head are as one fixed unit.

(3) To lead a target, the firer swings the rifle with a smooth, uniform motion by pivoting the body at the waist. There should be no independent movement of the arms, the shoulders, the head, or the rifle.

(4) The instructor explains and demonstrates the position, and points out that if the rifle is pulled or pushed in the desired direction by means of the left hand and arm, the rifle will move with a jerky motion, thereby increasing the possibility of jerking the trigger; or the front sight may be pulled or pushed out of alinement with the rear sight and the eye.

(5) Position exercises are conducted until the soldier becomes proficient in assuming positions rapidly for firing at hostile aircraft moving in any direction.

■ 85. SECOND STEP: AIMING AND LEADING EXERCISES.—*a. General.*—(1) *Purpose.*—The purpose of the aiming and leading exercises is to teach the correct method of aiming and to develop skill in swinging the rifle with a smooth, uniform motion so as to maintain the aim on aerial targets.

(2) *Method.*—(*a*) Pupils assigned to the nonoverhead targets take up the standing ready position in one line at about 1½-yard intervals, 500 inches from and facing the assigned target. Coaches take positions that enable them to observe their pupils. The commands for the exercise are: 1. AIMING AND LEADING EXERCISE, 2. ONE (TWO OR THREE) TARGET-LENGTH LEAD (S), 3. TARGETS. At the command TARGETS, the targets are operated at a speed of from 15 to 20 feet per second; the pupils assume the firing position rapidly, aline the sights on the spotter indicating the proper lead, and take up the slack in the trigger; then they swing the rifle with a smooth, uniform motion by pivoting the body at the waist, and maintain the aim on the proper lead during the travel of the target. the operation is repeated as the target is moved in the opposite direction. The exercise is continued until the target has

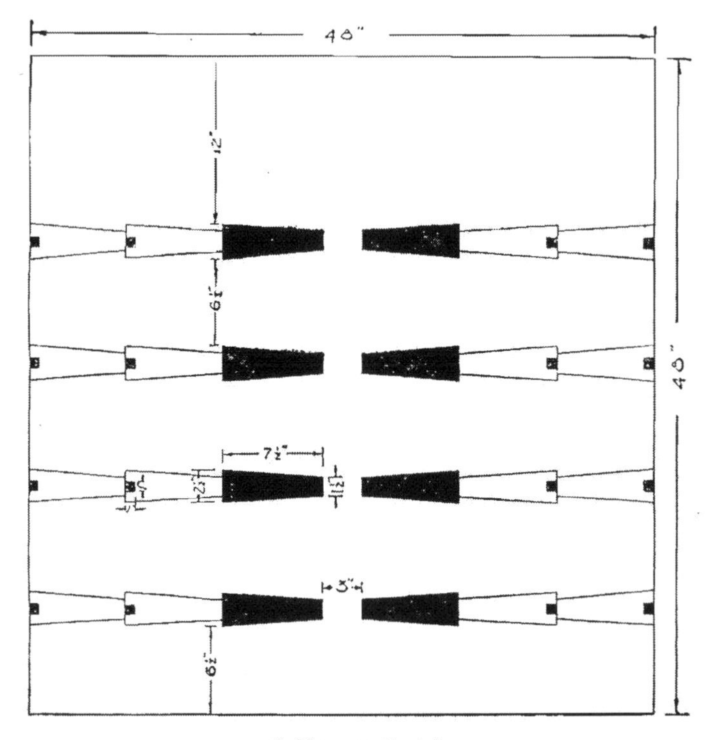

① Nonoverhead.

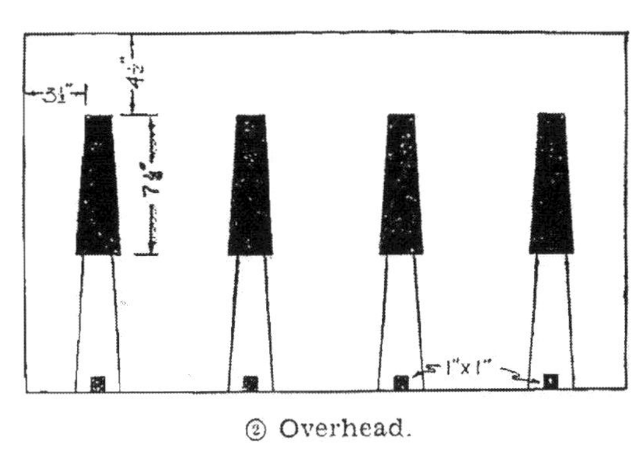

② Overhead.

FIGURE 36.—Aiming and leading targets.

been moved five times in each direction. The coach and pupil then change places, and the exercise is continued until all men have acquired some skill in aiming and leading with one, two, and three target-length leads, both from right to left and left to right.

(*b*) For the group assigned to the overhead target, the line is formed perpendicular to and facing the line of flight of the target. The procedure is the same except that one or two target-length leads only are used. (See fig. 36 for targets used.)

(3) *Importance of correct position.*—The importance of correct position and of swinging the rifle with a smooth, uniform motion by pivoting the body at the waist should be constantly emphasized.

b. Duties of coach.—In this exercise the coach sees that the—

(1) Proper position is taken.

(2) Slack is taken up promptly and firmly.

(3) Rifle is swung with a smooth motion.

(4) Rifle is swung by pivoting the body at the waist.

(5) Arms, shoulders, rifle, and head move as a unit with the rifle.

■ 86. THIRD STEP: TRIGGER-SQUEEZE EXERCISES.—*a. Importance.*—(1) Correct trigger squeeze is the most important operation to be performed in firing the rifle. The rifleman is trained to squeeze the trigger exactly as he does in rapid fire at stationary ground targets except that the rifle is kept in motion during the trigger squeeze, the firing of the shot, and momentarily after the shot is fired.

(2) In firing at a rapidly moving target the untrained man has a tendency to permit the rifle to come to rest momentarily while applying the final squeeze. This causes the shot to pass behind the target. Another tendency is to pull the trigger quickly the instant the aim is on the required lead. This causes flinching.

(3) Unless men are trained to apply pressure on the trigger in such a way that they cannot know the exact instant the cartridge will be discharged, all other training is a waste of time.

(4) Owing to the short period of time during which the usual aerial target is within effective range, fire should be opened as soon as possible and delivered as rapidly as accuracy permits. The trigger must be squeezed aggressively

and decisively. Once started, the pressure is continued until the cartridge is fired.

(5) Skill in squeezing the trigger properly when firing at rapidly moving targets is difficult to acquire. Although men will have had training in trigger squeeze during their course in stationary target marksmanship, firing at rapidly moving targets introduces certain additional elements which must be overcome before skill is acquired. The greater part of the time allotted to preparatory exercises should therefore be devoted to trigger-squeeze exercises.

b. Object.—The object of the trigger-squeeze exercises is to train the rifleman to apply pressure on the trigger while keeping the rifle in motion, to develop a decisive trigger squeeze so that fire can be opened in a minimum of time without loss of accuracy, and to train him to follow through with the shot.

c. Method.—(1) Trigger-squeeze exercises are conducted in a manner similar to the aiming and leading exercises. The targets used are also the same except that the spotters indicating the lead are removed. (See fig. 37.) If the spotters indicating the lead are left on the target they will cause an increased tendency of the pupil to pull the trigger quickly the instant the aim is on the spotter, thereby defeating the purpose of the exercises.

(2) The pupils in the standing ready position are placed in one line at about 1½-yard intervals, 500 inches from and facing the assigned nonoverhead target. The coaches take position so they can observe their pupils. The commands for the exercise are: 1. SIMULATE LOAD, 2. TRIGGER-SQUEEZE EXERCISE, 3. ONE (TWO, OR THREE) TARGET-LENGTH LEAD(S). 4. TARGETS. At the command TARGETS, the targets are operated at the proper speed; the pupils rapidly assume the firing position; take up the slack in the trigger; apply the target length in measuring the lead announced in the order; direct the aim on that point; maintain the aim at the proper lead (by swinging the rifle in the manner taught) at the same time applying a constantly increasing pressure on the trigger until the striker is released. The aim and pressure on the trigger are maintained during the entire length of travel of the target regardless of the time of release of the

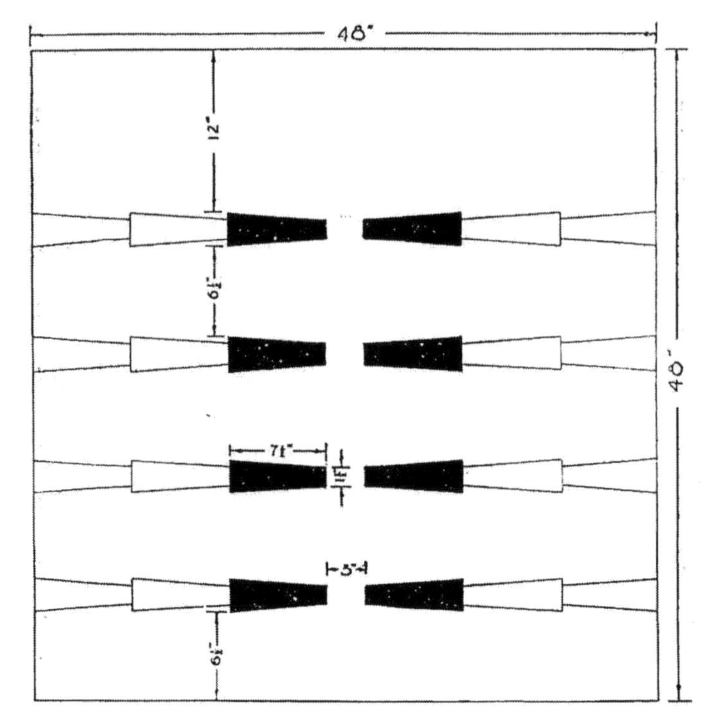

① Nonoverhead.

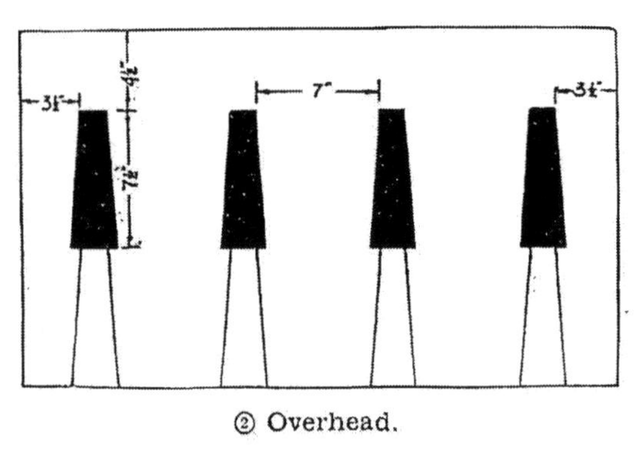

② Overhead.

FIGURE 37.—Instruction targets.

striker. The importance of following through with the shot must be strongly emphasized. It is only by following through that men develop the habit of keeping their rifle in motion during the entire process of firing. All of these steps are performed as one continuous operation. The exercise consists of five passages of the target in each direction. The coach and pupil then change places. The work is continued until all men have become proficient.

(3) The procedure for the overhead trigger-squeeze exercise is the same except that the line is formed perpendicular to and facing the flight of the target, and only one or two target-length leads are used.

d. Duties of coach.—In the trigger-squeeze exercise the coach sees that the—

(1) Proper position is taken.

(2) Slack is taken up promptly and firmly.

(3) Rifle is swung with a smooth, uniform motion.

(4) Rifle is swung by pivoting the body at the waist.

(5) Arms, shoulders, rifle, and head move as a unit as the rifle is swung.

(6) Pressure on the trigger is applied promptly, decisively, and continuously.

(7) Eye is kept open and does not blink at the instant the striker falls.

(8) Muzzle does not jerk coincident with the release of the striker.

(9) Pupil maintains the aim and trigger pressure during the entire length of travel of the target.

SECTION IV

MINIATURE RANGE PRACTICE

■ 87. GENERAL.—*a.* Miniature range practice is divided into two parts: instruction firing and group firing. There is no record firing.

b. All firing is on moving targets on the 500-inch range. A suggested arrangement of the range is given in paragraph 97. Provision is made for simultaneous firing by separate groups on the horizontal, the diving, the climbing, and the overhead targets.

c. The course should first be fired with the caliber .22 rifle after which, if ammunition and danger area will permit, the caliber .30 rifle may be used.

d. All rifles are zeroed before range practice starts.

■ 88. SAFETY PRECAUTIONS.—*a.* Bolts of rifles are kept open at all times except when men are firing or simulating fire on the firing line.

b. Rifles are loaded only at the command of the officer in charge of the firing on each target.

c. At the completion of the firing of a score all rifles are unloaded and bolts opened.

d. If firers go forward to inspect their targets, rifles are left on the firing line with bolts open.

e. No one is allowed in advance of the firing line except by permission of the officer in charge of that particular target. The officers in question do not grant such permission until they have assured themselves that all rifles are unloaded and the bolts are open.

f. Target operators remain behind the protective wall except when ordered to leave by the officer in charge of the target which they are operating.

g. Before permitting any firing, the officer in charge of each target instructs all members of a group that the bolt of the caliber .22 rifle must not be forced home if difficulty in feeding occurs. An attempt to force the bolt home in such circumstances may ignite the cartridge before it is chambered.

■ 89. INSTRUCTION FIRING.—*a. General.*—(1) The purpose of instruction firing is to practice in actual firing the principles taught in the preparatory exercises.

(2) During instruction firing the soldier works under the supervision of a coach.

(3) As soon as a group completes its preparatory training on a target, it fires its instruction practice on that target before moving on to the next target.

(4) Instruction firing is covered in table I below.

b. Procedure.—(1) Since the instruction firing on each type of target follows immediately after the preparatory exercises on that target, the organization of the unit for firing should be the same as that given in paragraph 83*b*.

(2) The front rank of each group is formed on the firing line *in the kneeling firing position.* Men in the rear rank act as coaches.

(3) One-half of the front rank fires while the remaining front rank men simulate firing.

(4) Silhouettes are assigned to each firer. For example, the four silhouettes on the right of the targets are assigned the first four men on the right of the line; the four silhou-

ettes on the left of the targets are assigned the next four men. Silhouettes for the men simulating firing are assigned in the same manner, that is, the right four are assigned silhouettes on the right of the target and the left four are assigned silhouettes on the left of the target.

(5) The officer in charge of the target commands: 1. LOAD, 2. ONE (TWO-THREE) TARGET-LENGTH LEAD(S), 3. TARGETS. At the command TARGETS, the targets are operated at the proper speed. Men assigned silhouettes on the right half of the nonoverhead targets mentally apply the target length in measuring the lead announced. They direct their aim on that point and, while maintaining the aim, squeeze the trigger until the shot is fired. They continue to maintain the aim during the entire length of travel of the target regardless of the time the shot was fired. They fire one shot each time the target crosses from their left to right. The men assigned silhouettes on the left half of the same targets aim and fire one shot in the same manner as explained above each time the target crosses from their right to left.

(6) Men assigned silhouettes on the overhead target fire one round each time the target is run in the approaching direction in exactly the same manner as explained above.

(7) Five rounds constitute a score. After each string of five rounds, targets are scored and shot holes penciled.

(8) One point is awarded for each hit in the silhouette when using one target-length lead or in the proper scoring space when using more than one target-length lead.

(9) Half-groups alternate firing and simulating firing.

(10) When front rank men have fired two scores, one score as the target moved in each direction, they change places with the men in the rear rank. They coach the rear rank men who become the firers.

(11) This procedure is followed until all men of the group have performed the required firing at that target.

(12) Upon completion of the firing prescribed in table I for any one type of target, the group moves to another type target and continues until all have completed the instruction firing.

(13) Modifications of the foregoing method of firing to meet local conditions are authorized.

TABLE I.—*Instruction firing*

[Range 500 inches]

Target	1 lead, 10 rounds	2 leads, 10 rounds	3 leads, 10 rounds	Total
Horizontal	5 rounds R to L 5 rounds L to R	5 rounds R to L 5 rounds L to R	5 rounds R to L 5 rounds L to R	30
Climbing	5 rounds R to L 5 rounds L to R	5 rounds R to L 5 rounds L to R	5 rounds R to L 5 rounds L to R	30
Diving	5 rounds R to L 5 rounds L to R	5 rounds R to L 5 rounds L to R	5 rounds R to L 5 rounds L to R	30
Overhead	5 rounds approaching. 5 rounds receding	5 rounds approaching. 5 rounds receding		20

NOTE.—Speed of all targets, 15 to 20 feet per second. Total rounds, 110.

■ 90. GROUP FIRING.—*a. General.*—Group firing is the final phase of antiaircraft marksmanship training on the miniature range. It provides for competitions and illustrates the effectiveness of the combined fire of a number of riflemen. It is not started until preparatory training and instruction firing have been completed.

b. Procedure.—(1) Two silhouettes are assigned to each squad or similar group. One is fired on as the target moves from left to right; the other as the target moves from right to left.

(2) Each man in the front rank, then each man in the rear rank, fires five rounds at each silhouette as the target moves in the appropriate direction.

(3) Targets are not scored until completion of the firing of the entire squad or group.

c. Scoring.—A value of 1 is given each hit on the silhouette.

SECTION V

TOWED-TARGET FIRING

■ 91. GENERAL.—*a.* Towed-target firing is the final phase of antiaircraft rifle marksmanship. It is conducted on the towed-target range described in paragraph 98.

b. It consists of firing at a sleeve target at various ranges and on varied courses with caliber .30 ball or tracer ammunition.

c. The towed-target courses described in this manual are guides; they may be modified as required. Safety measures and ammunition requirements restrict the length of the course. Safety measures also prevent the adoption of courses such as those on which the target, moving at a low altitude, is receding from or diving at the firing line.

d. Towed-target firing follows miniature range firing. If, owing to lack of facilities, a unit is unable to conduct miniature range firing it may be permitted to conduct towed-target firing provided antiaircraft marksmanship preparatory training has been completed.

■ 92. COURSES TO BE FIRED.—Units authorized to fire will fire one or more of the courses enumerated in table II.

TABLE II.—*Courses to be fired*

Course No.	Type of flight	Altitude of target	Horizontal range of course (yards) [1]	Speed	Remarks
1	Nonoverhead—horizontal (parallel to firing line from left to right).	Minimum consistent with safety.	Minimum 100; maximum depends on width of danger area of range.	Maximum possible.	See fig. 48.
2	Nonoverhead—horizontal (parallel to firing line from right to left).	do	do	do	Do.
3	Overhead (perpendicular to firing line).	do	Minimum 100; maximum 450.	do	See fig. 49.
4	Combine courses 1, 2, and 3.	do	Same as for courses 1, 2, and 3.	do	See fig. 50.

[1] The horizontal distance from the firing point to a point directly under the target. The maximum slant range for all courses should not exceed 600 yards.

■ 93. SAFETY PRECAUTIONS.—*a*. Towed-target firing is conducted with due regard for the safety of the pilot of the towing airplane, the personnel engaged in firing, and spectators.

b. All firing must be controlled by suitable signals or commands. COMMENCE FIRING and CEASE FIRING must be given in such a manner as to be understood clearly and quickly by everyone engaged in firing.

c. The signals and commands for COMMENCE FIRING and CEASE FIRING are given at such time as to prevent any bullets from falling outside the danger area.

d. For all overhead flights, the signal or command for COMMENCE FIRING is not given until the towing airplane has reached a point 50 yards or less from the firing line (measured horizontally on the ground) and there is no danger of bullets striking the airplane. The signal or command for CEASE FIRING is given before the sleeve target reaches a point 100 yards in advance of the firing line (measured horizontally on the ground), so there is no danger of bullets dropping outside the firing area.

e. Whenever a towing cable breaks and the towing airplane is on a course which passes near the firing line, all personnel in the vicinity are warned to lie flat on the ground until danger from the loose cable and the release is past.

f. In no circumstances will rifles be pointed at or near the towing airplane. All tracking will be on the towed target. Muzzles will be depressed during loading.

g. At least two safety officers are designated to assist the officer in charge of firing in carrying out safety precautions.

h. Firing is permitted only when the angle between the gun-target line and the tow line is greater than 45°.

i. An officer of the Army Air Forces should be at the firing point during an organization's initial practice for the season to check the safety measures taken.

j. Additional safety precautions are covered in AR 750–10.

■ 94. FIRING PROCEDURE.—*a.* The men to fire take position on the firing line with at least 1½-yard intervals.

b. The officer in charge of firing takes position in rear of the center of the firing line.

c. Safety officers take position at both flanks of the firing line.

d. As the *towing airplane* approaches the left (right) side of the danger area, the officer in charge of firing gives the command: 1. (SO MANY) ROUNDS, LOAD, 2. SLEEVE TARGET APPROACHING FROM THE LEFT (RIGHT). Each rifleman loads and locks his piece.

e. As the towed target approaches the danger area, the officer in charge of firing commands: 3. (So many) TARGET-LENGTH LEADS. At this command, each rifleman unlocks his piece, aims by swinging through the sleeve to the announced lead, pivoting at the waist, and maintains his estimated lead.

f. In firing at crossing targets the safety officer who is stationed at the end of the firing line opposite the target's approach signals or commands: COMMENCE FIRING, when the sleeve target has completely crossed the line marking the firing area. The officer in charge of firing and his assistants repeat the command or signal to insure that all firers hear it. Each rifleman squeezes the trigger until the first shot is fired. He continues to reload, re-aim, and fire until the command or signal CEASE FIRING is given. The safety officer at the end of the firing point opposite to the target's departure watches the flight of the sleeve target during the firing. When he sees that the sleeve is about to leave the danger area he signals or commands: CEASE FIRING. The officer in charge of firing and his assistants repeat the command or signal to insure that all firers hear it.

g. In firing at overhead targets, the same procedure is followed except that the officer in charge of firing, from his position behind the center of the firing line, determines when firing commences and ceases. He gives the command or signal to COMMENCE FIRING when the towing airplane is 50 yards or less in advance of the firing line, and the command CEASE FIRING before the sleeve is 100 yards in advance of the firing line. (See par. 93*d*.)

■ 95. SCORING.—*a*. The number of hits is found by dividing the number of holes in the target by two. An odd hole is counted as a hit.

b. The hit percentage is obtained by dividing the number of hits as obtained in *a* above by the total number of rounds fired at the target.

SECTION VI

RANGES, TARGETS, AND EQUIPMENT

■ 96. RANGE OFFICER.—A range officer is appointed well in advance of range practice. His chief duties are—

a. To make timely estimates for material and labor to place the range in proper condition for firing.

b. To supervise and direct repairs and alterations to installations.

c. To instruct and supervise range guards.

■ 97. MINIATURE RANGE.—*a*. The miniature range consists of—

1 horizontal target (fig. 38).

1 double climbing and diving target (fig. 39).

1 overhead target (fig. 40).

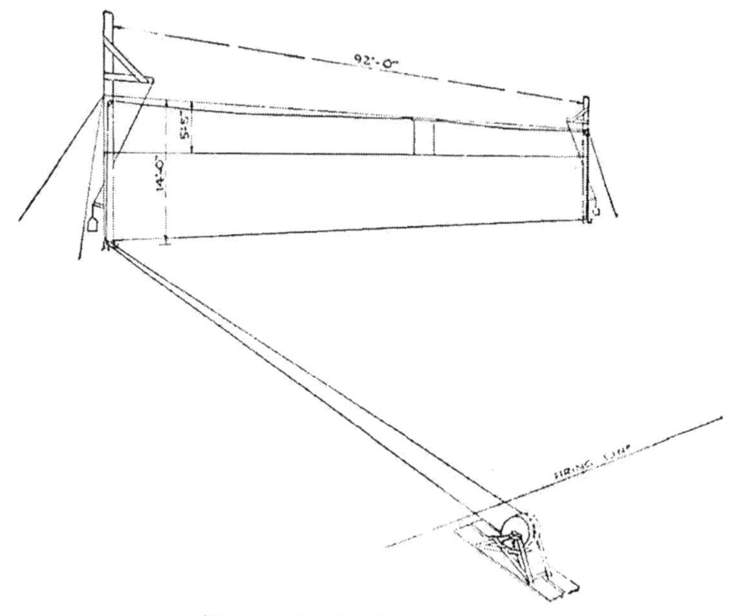

FIGURE 38.—Horizontal target.

b. A suggested arrangement of the targets is shown in figure 41.

c. For details of range apparatus see figures 42 to 46, inclusive.

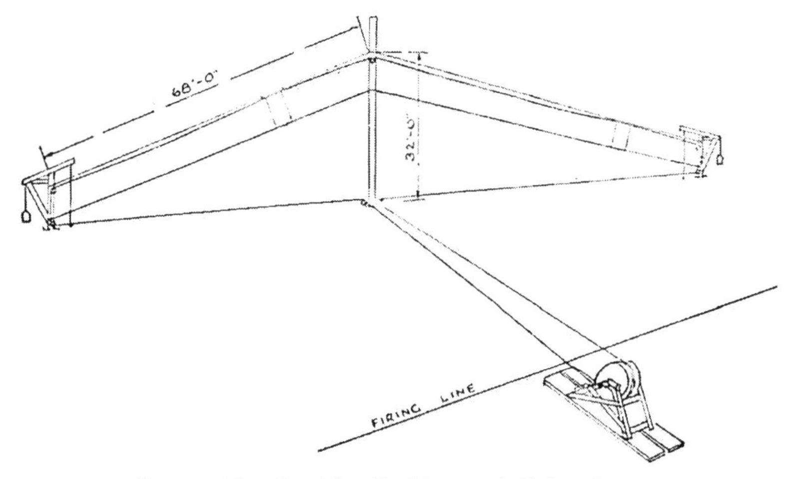

FIGURE 39.—Double climbing and diving target.

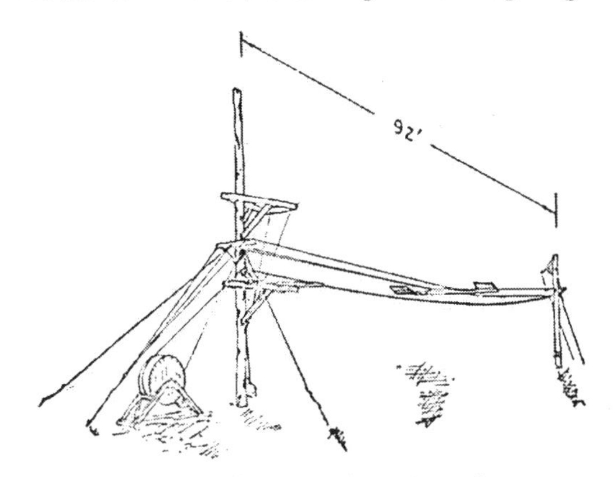

FIGURE 40.—Overhead target.

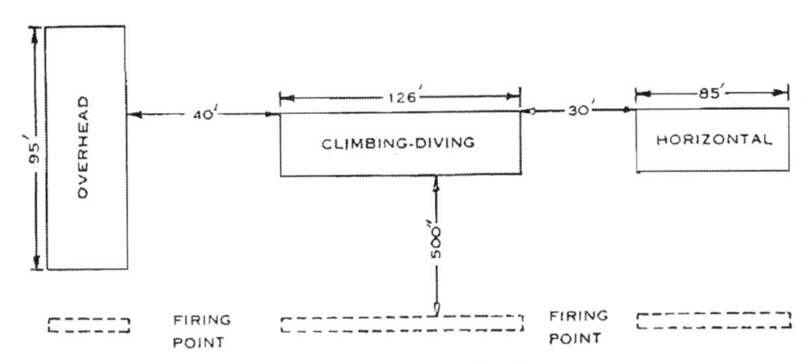

FIGURE 41.—Arrangement of targets.

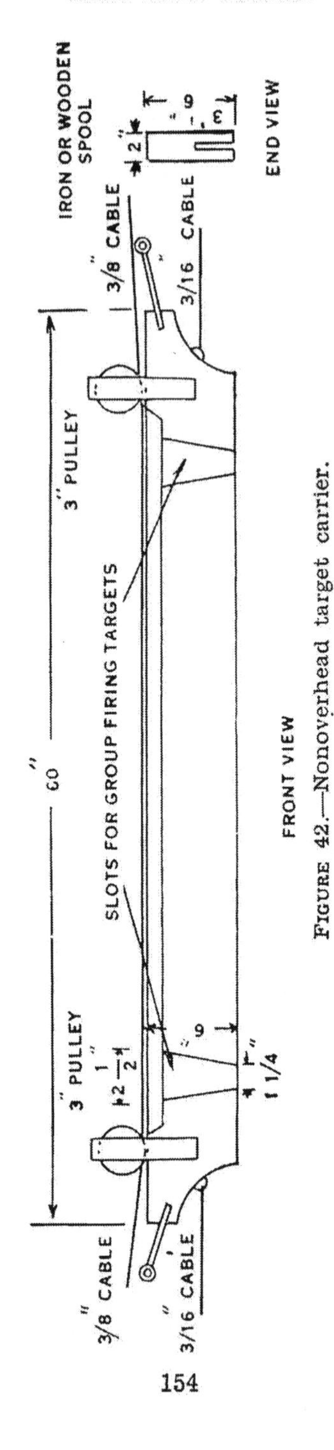

FIGURE 42.—Nonoverhead target carrier.

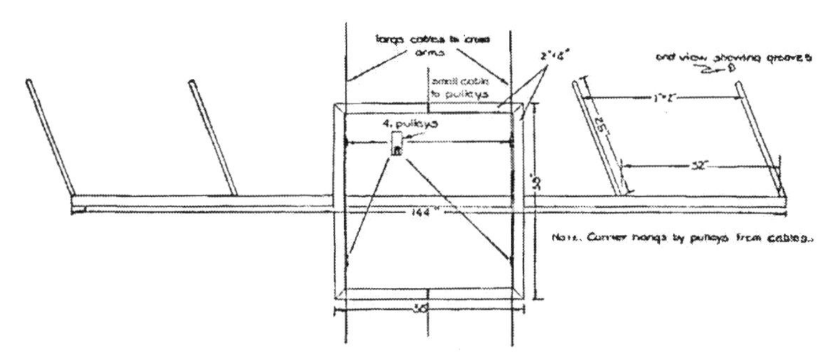

FIGURE 43.—Overhead target carrier.

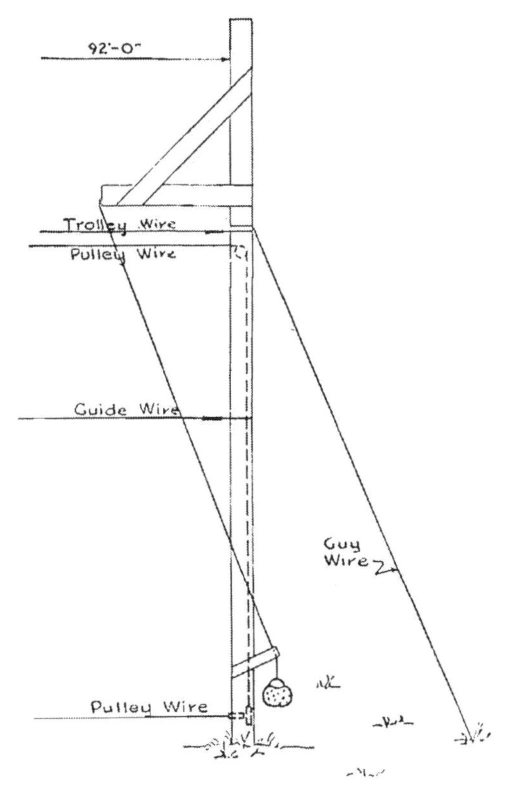

FIGURE 44.—Bumper and wire arrangement.

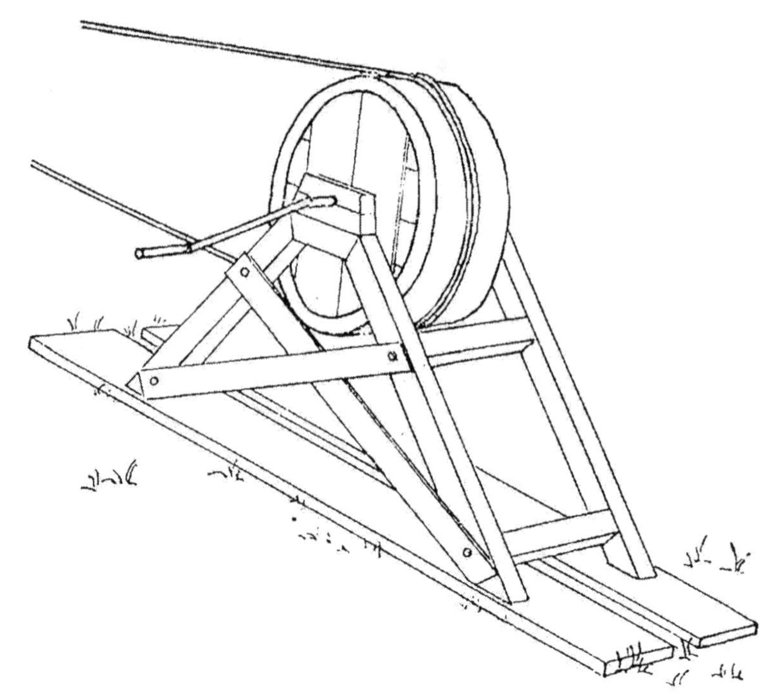

FIGURE 45.—Operating drum.

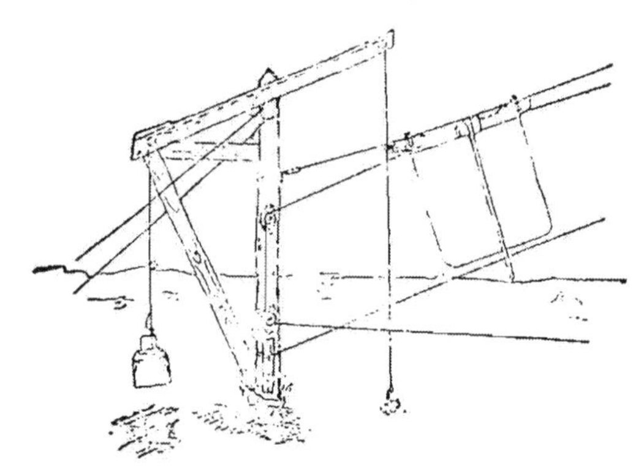

FIGURE 46.—Rear view of climbing and diving target and method of securing target to frame.

d. Danger area required.—(1) The danger area required depends upon the type of ammunition used. (See AR 750–10 for size and shape.)

(2) The miniature range may be laid out as described in paragraph 98. Care must be taken to place the firing line and targets so that no fire will fall outside of the danger area.

e. Equipment required.—If the organization for training suggested in paragraph 83*b* is used, the following equipment is necessary:

64 caliber .22 rifles (if available).
4 aiming and leading targets (see fig. 36) made of beaverboard with silhouettes pasted on.
6 instruction firing targets per range (see fig. 37). (Same as aiming and leading targets except that spotters are eliminated.)
1 score card per man as follows:

INDIVIDUAL SCORE CARD

ANTIAIRCRAFT RIFLE MARKSMANSHIP

Date__________________, 19____

Name____________________________

Target	1 TL lead			2 TL lead			3 TL lead		
	Rounds fired	Hits	Per-cent	Rounds fired	Hits	Per-cent	Rounds fired	Hits	Per-cent
Horizontal									
Climbing									
Diving									
Overhead									
Total									

■ 98. TOWED-TARGET RANGE.—*a.* In selecting the location of a towed-target range the danger area is the chief consideration. (See AR 750-10.)

b. The firing point should accommodate at least 50 men in line with a 1½-yard interval between men. A level strip of ground, preferably on a hill, 75 yards long and 2 yards wide is suitable. A firing point similar to the firing point of a class A rifle range may be built.

c. (1) After the towed-target range has been selected, the firing point, limits of fire, and danger area should be plotted on a map or sketch of the area.

(2) From this map or sketch the range is then laid out on the ground. First, each end of the firing point is marked by a large stake. The right and left limits of fire are then marked by posts. Each post is placed at the maximum distance at which it is plainly visible from the firing point. When these distances have been determined, the posts are located in azimuth by the following method: To locate the post marking the left limit of fire, an aiming circle or other

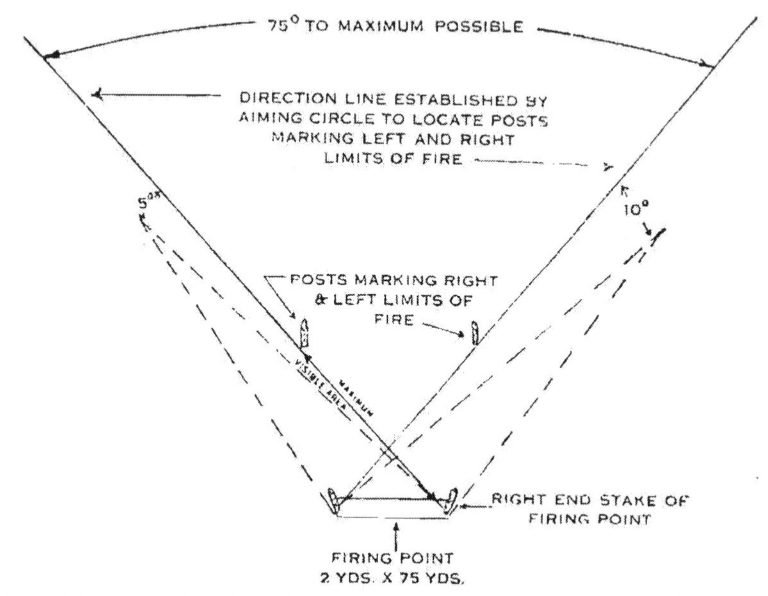

FIGURE 47.—Towed-target range showing firing point and limits of fire. Dotted lines show danger area.

angle-measuring instrument is set up at the right end stake of the firing point. It is then oriented and laid on an azimuth which, by reference to the map or sketch, is known to be the farthest to the left that the rifle at the right end of the firing point can safely be fired. The post marking the right limit of fire is located in a similar manner with the instrument set up at the left end stake of the firing point. (See fig. 47.)

(3) Direction guides for the towing airplane to follow should be distinctly marked on the ground for each course within the limits of fire. White targets or strips of cloth placed flat on the ground about 30 feet apart are suitable.

■ 99. TOWED TARGETS.—*a. Type and source.*—The targets used in towed-target firing are sleeve targets furnished by the unit of the Army Air Forces assigned the towing mission. They are returned to that unit after they have been scored.

b. Towline.—The towing line will not be less than 600 yards long.

■ 100. INSTRUCTIONS TO PILOTS FOR TOWING MISSIONS.—*a.* Towed-target firing requires the closest cooperation between the pilot of the towing airplane and the officer in charge of firing. Decisions affecting the safety of the airplane rest with Army Air Forces personnel.

b. The air mission for towed-target firing will be specifically stated. The commanding officer requesting airplanes for towed-target firing will furnish, in writing, to the Army Air Forces unit commander concerned the following information:

(1) Place of firing.

(2) Date and hour of firing.

(3) Number of missions to be flown, and altitude, course, speed, and number of runs for each.

(4) Ground signals to be used.

(5) Map of the area with firing line, angle of fire, danger area, course of each mission, and location of grounds for dropping targets and messages, all plotted. An alternate dropping ground is designated when practicable. Either or both dropping grounds are subject to approval of the pilot.

(6) Length of the towline, within limits established by the Army Air Forces, and subject to approval of the pilot.

(7) Number of sleeve targets required.

c. Whenever practicable, the officer in charge of the firing discusses with the pilot the detailed arrangements in *b* above. This discussion should take place on the towed-target range where the various range features can be pointed out to the pilot. The courses over which the airplane is to be flown should be distinguished on the ground (within the angle

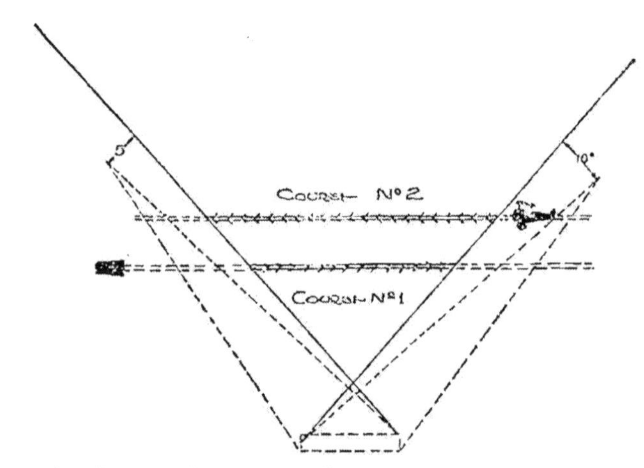

FIGURE 48.—Courses Nos. 1 and 2. Firing takes place when target is on shaded portion of course.

of fire). Machine-gun targets placed flat on the ground, about 30 feet apart, or strips of target cloth are practicable for this purpose on some courses. On others a terrain feature such as a clearly visible trail may be used.

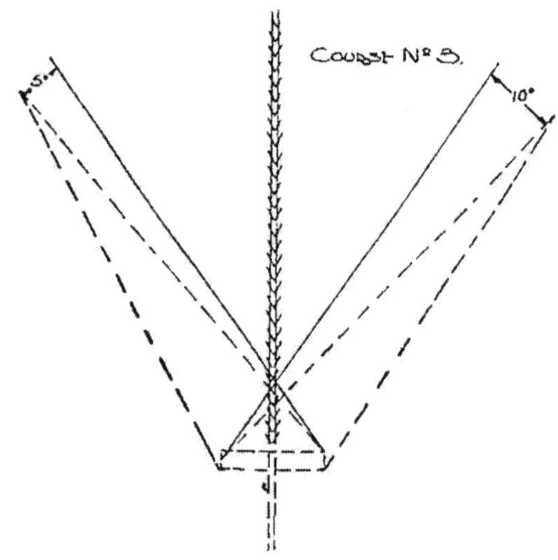

FIGURE 49.—Course No. 3. Firing takes place when target is on shaded portion. Fire is opened when towing airplane is 50 yards or less from firing point.

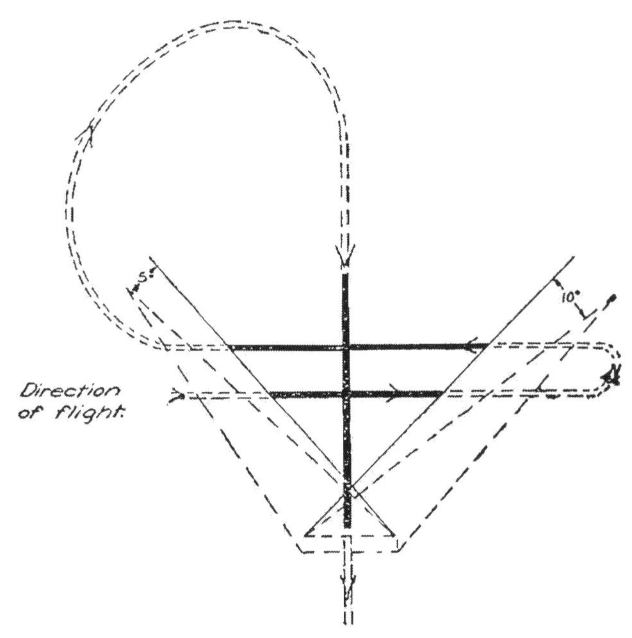

FIGURE 50.—Course No. 4. Heavy lines indicate when towed target is fired upon.

■ 101. SIGNALS.—*a*. Direct radio communication is the most effective means by which the officer in charge of towed-target firing and the pilot of the towing airplane maintain contact with each other. Panels should be available for use in the event radio communication fails.

b. For signaling from ground to air any method may be used that is agreed upon by the officer in charge of firing and the pilot of the towing airplane. Simple panel signals such as the following are useful:

Stand by	0 0 2
Ready fire	0 0 0
Repeat run No. 1	0 9 1
Repeat run No. 2	0 9 2
Repeat run No. 3	0 9 3
Repeat course	0 9 4
Mission complete	Pick up panels

c. The pilot may communicate with the officer in charge of firing by dropped messages, by rocking his wings, or by any other convenient method mutually agreed upon.

CHAPTER 5

TECHNIQUE OF RIFLE FIRE

SECTION I

GENERAL

■ 102. DEFINITIONS.—*a*. The training of riflemen for combat is progressive. It includes three phases. The *first phase* is individual training; it covers such allied subjects as rifle marksmanship, extended order, drill and combat signals, and certain elements of scouting and patrolling. The *second phase* covers the technique of fire. This is team training in the application and control of the collective fire of rifle fire units. In the *third phase* the individual and team training of the first and second phases are combined with tactical training; this phase is called combat firing. This chapter deals with the second phase of training.

b. *Collective fire* is the combined fire of a group of individuals.

c. A *fire unit* is a unit whose fire in battle is under the immediate and effective control of its leader. The usual rifle fire unit is the rifle squad.

■ 103. IMPORTANCE OF RIFLE FIRE.—Effective rifle fire is a characteristic of successful Infantry. It is an element which may determine the issue of battle. Collective fire is most effective when it is the product of teamwork. Training in the technique of fire, as set forth in this chapter, is designed to train rifle squads to act as efficient and dependable teams in the application and control of collective fire.

■ 104. PROGRESSIVE STEPS.—*a.* Although training in the technique of rifle fire is conducted without regard to tactical considerations, its application to tactical situations should be kept in mind. The instruction is progressive; it is divided into six consecutive steps. Each step includes some, or all, of the technique learned in previous steps. The six steps are—

(1) Range estimation.
(2) Target designation.
(3) Rifle fire and its effect.
(4) Application of fire.
(5) Landscape-target firing.
(6) Field-target firing.

b. A 13-week training schedule for mobilization should include about 30 hours for this instruction. About 20 hours should be devoted to the first five steps and 10 hours to the sixth step. (See MTP 7–3.)

c. It is not essential that perfection be reached in each step before proceeding to the next step; it is better to reach the required standard by repeating in succeeding steps all that has been learned before. It is necessary, however, that each step be understood before proceeding to the next. The instructor explains each step in detail. He then makes plain its relation to the subject as a whole. This is followed by a demonstration which in turn is explained by the instructor. The men then practice the principles and methods that have been explained and demonstrated. Exercises pertaining to the various steps are described in detail in this chapter. They can be used for demonstrations and instruction practice. Some of them can be used to test fire units, thus introducing the desirable element of competition. These exercises may be changed to conform to local conditions.

SECTION II

RANGE ESTIMATION

■ 105. IMPORTANCE.—In battle, ranges are seldom known in advance. Therefore, the effectiveness of rifle fire depends in large measure upon the ability of riflemen to estimate ranges quickly and accurately.

■ 106. METHODS.—Only three methods of estimating ranges are taught in the technique of rifle fire. These are—

Use of tracer bullets.
Observation of fire.
Estimation by eye.

■ 107. USE OF TRACER BULLETS.—When the range to a target is being estimated by the use of tracer bullets, the scout or leader first estimates the range by eye, fires a tracer bullet, corrects the sight-setting according to the strike of the bullet, and continues to fire and correct the sight-setting until a tracer appears to strike the target. The estimator then announces the correct range after due consideration of the zero of his rifle.

■ 108. OBSERVATION OF FIRE.—This method can be used when the ground is dry and the strike of the bullets can be seen. The same procedure is followed as in determining the range by tracer bullets. The following points must be taken into consideration:

a. Dust will appear slightly above the striking point of the bullet.

b. As seen from the firing point, dust appears on the side toward which the wind is blowing. Thus, if the wind is blowing from right to left, dust will appear to the left of the point of impact.

c. As seen from a point on the flank, shots that hit beyond the target appear to strike on the side on which the observer is posted; those that fall short appear to strike on the side away from the observer.

■ 109. ESTIMATION BY EYE.—*a. Necessity for training.*—In combat, ranges are usually estimated by eye. Untrained men make an average error of 15 percent of the range when estimating by eye. Hence a definite system of range estimation, coupled with frequent practice on varied terrain, is essential to accuracy with this method.

b. Unit-of-measure method.—(1) Ranges less than 500 yards are measured by applying a mental unit of measure 100 yards long. Thorough familiarity with the 100-yard unit, and with its appearance on varied terrain and at different distances, is necessary if the soldier is to apply it accurately.

(2) Ranges greater than 500 yards are estimated by selecting a point halfway to the target, applying the unit of measure to this halfway point, and doubling the result.

(3) The average of a number of estimates by different men is usually more accurate than a single estimate. This procedure may be used when time permits.

c. Appearance of objects.—In some situations much of the ground between the observer and the target is hidden from view, which makes the application of the unit of measure impossible. In that event the range is estimated by the appearance of objects. Whenever the appearance of objects is used as a basis for range estimation, the observer must make allowance for the following effects:

(1) Objects seem nearer—

(*a*) When the object is in a bright light.

(*b*) When the color of the object contrasts sharply with the color of the background.

(*c*) When the observer is looking over water, snow, or a uniform surface such as a wheat field.

(*d*) When the observer is looking downward from a height.

(*e*) In the clear atmosphere of high altitudes.

(*f*) When the observer is looking over a depression most of which is hidden.

(2) Objects seem more distant—

(*a*) When the observer is looking over a depression most of which is visible.

(*b*) When there is a poor light or fog.

(*c*) When only a small part of the object can be seen.

(*d*) When the observer is looking from low ground toward higher ground.

d. Exercises.—(1) *No. 1.*—(*a*) *Purpose.*—To familiarize the soldier with the 100-yard unit of measure.

(*b*) *Method.*—The 100-yard unit of measure is previously staked out over varied ground with markers that are visible up to 500 yards. The men are required to become thoroughly familiar with the appearance of the unit of measure from the prone, kneeling, and standing positions at various ranges. They do this by studying the appearance of the unit from distances of 100, 200, 300, and 400 yards.

(2) *No. 2.*—(*a*) *Purpose.*—To illustrate the application of the unit of measure.

(*b*) *Method.*

1. Ranges up to 900 yards are measured accurately and marked at every 100 yards by large markers or target frames, each bearing a number to indicate its range. Men undergoing instruction are then placed about 25 yards to one side of the prolonged line of markers and directed to hold a hat or other object as an eye cover to exclude from view all of the markers. They are then directed to apply the unit of measure five times along a straight line parallel to the line of markers. When they have selected the final point the eye cover is removed, and the estimations of the successive 100-yard points and the final point are checked against the markers. Accuracy is gained by repeating the exercise.

2. Ranges greater than 500 yards are then considered. With the markers concealed from view, men estimate the ranges to points which are obviously more than 500 yards away and a little to one side of the line of markers. As soon as they have announced each range they remove their eye covers and check the range to the target and to the halfway point by means of the markers. Prone, sitting or kneeling, and standing positions are used during this exercise.

(3) *No. 3.*—(*a*) *Purpose.*—To give practice in range estimation.

(*b*) *Method.*—From a suitable point, ranges are previously measured to objects within 1,000 yards. The men are required to estimate and write down the ranges to various objects pointed out by the instructor. At least one-half of the estimates are made from the prone or sitting positions. Thirty seconds are allowed for each estimate. When all ranges have been estimated the papers are collected and the true ranges announced to the class. To create interest, individual estimates and squad averages may be posted on bulletin boards.

SECTION III

TARGET DESIGNATION

■ 110. IMPORTANCE.—Target designation is a vital element in the technique of rifle fire. Battlefield targets are usually so indistinct that leaders and troops must be able to designate their location and extent. Enemy troops are ordinarily so well-concealed that individual enemy soldiers are not visible. To cover such a target, squad leaders must be able to designate the area in which hostile troops are located, and members of the squad must be trained to place a heavy fire on the designated area even though no definite targets are visible.

■ 111. INSTRUCTION.—Prior to instruction in target designation, riflemen should understand those topographical terms most frequently used in designating targets; for example, crest, military crest, hill, cut, fill, ridge, bluff, ravine, crossroads, road junction, skyline, and the like.

■ 112. METHODS.—*a.* The following methods are used to designate targets:

(1) Tracer bullets.

(2) Pointing.

(3) Oral description.

(4) A combination of any two or all three of these methods.

b. Troops are trained in *all* the methods of target designation (pars. 113 to 115, incl.). In any given situation that method which is the fastest, simplest, and clearest is the best.

■ 113. TRACER BULLETS.—*a.* The use of tracer bullets is a quick and sure method of designating an obscure battlefield target. Tracers furnish the most accurate means of indicating the flanks of such a target. Their use is invariable when scouts or other members of the squad are already under fire; when the squad is deployed and separated, and out of voice range of the leader; or when cover is scarce, and pointing or other movement will expose personnel to hostile fire. On the other hand, tracers may disclose the position of the firer to the enemy. Furthermore, the surprise effect of a sudden burst of fire may be lost if it is preceded by tracers.

b. To designate a point target by this method, the scout or leader announces "Range 500, watch my tracer," and fires a tracer at the target. The flanks of a linear target are indicated by firing a tracer at each flank and announcing each shot as "Left flank," or "Right flank." Any range correction obtained by tracer firing is anncunced.

■ 114. Pointing.—Targets may be pointed out by the arm or the rifle. Pointing may be supplemented by oral description. When the rifle is used, it is canted to the right and aimed at the target. The head is then straightened up, care being taken not to move the rifle. A soldier standing behind looks through the sights and locates the target. If time permits, a bayonet can be stuck in the ground as a rest for a rifle aimed at the target. The range is always announced. Usually some supplementary description is necessary.

■ 115. Oral Description.—*a. Use.*—Oral description is often used by leaders to designate targets to their units. Battlefield conditions rarely permit a leader to designate a target direct to all members of his unit by this method. For this reason he usually combines tracers or pointing with his oral description.

b. Elements of oral target designation.—The elements of oral target designation are—

Range.
Direction.
Description of target.

These elements are always given in the above sequence with a slight pause between each element. An exception to this rule occurs when the target is expected to be visible for a short time only. In this event it is pointed out as quickly as possible. In such circumstances an oral target designation might be no more than "Those men." When the range is announced, men immediately set their sights before looking for the target.

c. Direction.—The terms *front* (left, right) and *flank* (left, right) may be used to indicate the general direction of the target. When necessary, the direction is fixed more accurately by methods described hereafter.

d. Simple description.—When the target is plainly visible, or at an easily recognized point, a simple description is used.

To designate a target located at A in figure 51, for example, this would be sufficient:

Range: 425.
Left front.
Sniper at base of dead tree.

e. Reference point.—(1) *General.*—When the target is indistinct or invisible and is not located at some prominent point, the direction of the target is indicated by the use of a reference point. This is usually some prominent, easily identified object such as a lone tree, a conspicuous stump, a church steeple, a water tower or other clearly defined landmark by reference to which the location of other points may be determined. So far as practicable, leaders should not select reference points that may be confused with similar landmarks in the immediate vicinity. A reference point on a line with, but beyond the target gives greater accuracy than one between the observer and the target. For brevity the word *"reference"* rather than "reference point" is used in fire orders.

(2) *When reference point is on line with target.*—When the reference point is on line with the target, the description takes the following form (fig. 51, target at B):

Range: 450.
Reference: church spire.
Target: machine gun in edge of woods.

Note that the range announced is that to the target and not to the reference point. When the word "reference" is used the word "target" is also used to differentiate between the two. Another example follows (fig. 51, target at C):

Range: 350.
Left front.
Reference: black stump.
Target: sniper on left side of road.

(3) *When reference point is not on line with target.*—

(*a*) When the reference point is not on line with the target, it is necessary to indicate how far to the right or left of the reference point the target lies. This distance is measured in terms of "fingers" or "sights" (see par. 116*b*). Suppose the hand is held so that the left edge of the fore finger is

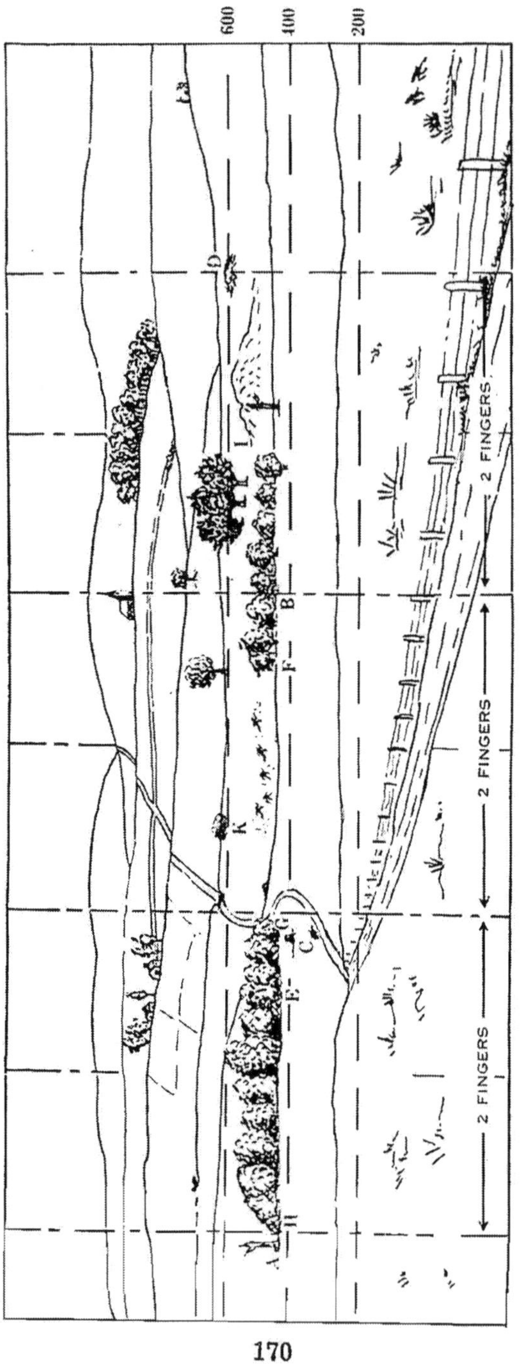

FIGURE 51.—Target designation.

on line with the reference point and it is found that the right edge of that same finger is in line with the target, the target is then one finger width to the right of the reference point and it is announced as "Right, one finger." If two fingers can be applied to the lateral interval between the reference point and target, the target is "Right, two fingers." If the sight leaf of the rifle is raised it may be used in the same manner as the finger (see par. 116*b*). If, for example, the target were found to be three sight widths to the left of the reference point, that fact would be announced as "Left, three sights." The following examples illustrate the use of "fingers" and "sights." (See fig. 51.)

(Target at D)—

Range: 600.

Reference: church spire. Right, two fingers.

Target: group of enemy in shell hole near crest.

(Target at E)—

Range: 425.

Left front.

Reference: dead tree. Right one and one-half fingers.

Target: sniper in edge of woods.

(Target at F)—

Range: 450.

Reference: church spire, left one-half sight.

Target: machine gun in corner of woods.

(*b*) The width or extent of targets may also be measured in fingers or sights (fig. 51).

(Target G to H)—

Range: 425.

Reference: church spire; left two fingers.

Target: enemy groups in edge of woods extending left two fingers.

(4) *Successive reference points.*—Successive reference points may be used instead of finger or sight measurements from one reference point (fig. 51).

(Target at I)—

Range: 500.

Reference: church spire; to the right and at a shorter range, group of three trees; to the right and at the same range.

Target: machine gun at left end of mound of earth.

(5) *Combinations.*—Combinations of successive reference points and finger or sight measurements may be used (fig. 51).

(Target at K)—

Range: 600.

Reference: church spire; to the left and at a shorter range, lone tree; left one sight and at the same range.

Target: machine gun in clump of brush.

f. Variations.—If one end of a target is considerably nearer than the other, the average range is announced, since dispersion will cover the target. Battlefield conditions impose many practical substitutions and combinations of methods in target designation. Frequently the squad leader is able to designate the target to only one or two members of his squad. Therefore, each member of the squad must be taught to assist in designating targets to other members of the squad team. At times the scouts designate the target to other members of the squad as they arrive in the vicinity. Formal, long-winded oral target designations confuse more often than they help.

■ 116. EXERCISES.—*a. No. 1.*—(1) *Purpose.*—To afford practice in target designation by means of tracer bullets.

(2) *Method.*—(*a*) On a class A or class B range a concealed target representing a machine gun is placed near a pit or other bulletproof shelter. About 500 yards in front of the target a firing position suitable for a squad is selected. The location of the target should be visible from the firing position, but the target itself may be completely concealed. This concealment for the target should be natural in order not to attract attention.

(*b*) The squad is deployed along the firing position. All except the scouts are then faced to the rear.

(*c*) The scouts take the prone position and are told that the waving of a red flag to their front will represent the firing and smoke from the machine gun.

(*d*) A man stationed in the pit waves a flag in front of the target for about 30 seconds and retires to the protection of the pit.

(*e*) The squad is faced to the front and men take the prone position. Rifles are loaded; the scouts load with tracer ammunition and the remainder of the squad with ball cartridges.

(*f*) The scouts point out the target by firing tracers and announcing the range, which is passed orally from man to man.

(*g*) As soon as each man understands the location of the target he opens fire with the proper sight setting.

(*h*) The instructor causes firing to cease shortly after all the men are firing.

(*i*) Noncommissioned officers do not participate in the fire. Squad leaders move about freely behind their men and observe the firing. The second in command assists the squad leader.

(*j*) After firing ceases, sight settings are checked by the squad leader and the target examined or the hits are signaled to the squad.

b. *No*. 2.—(1) *Purpose*.—To teach the use of fingers and sights for lateral measurement.

(2) *Method*.—(*a*) A number of short, vertical lines 1 foot apart are plainly marked on a wall or other vertical surface. At a distance of 20 feet from the wall a testing line is drawn or marked out by stakes. The instructor explains that the vertical lines are one finger (50 mils) or one sight (50 mils) apart when measured from the testing line, and are used to determine the correct distance the hand must be held from the eye for each finger to cover the 1-foot space between two of the vertical lines, and the point on the stock of the rifle where the eye must be placed for the raised sight leaf to cover the 1-foot space between two of the vertical lines.

(*b*) The instructor then explains and demonstrates the use of fingers in lateral measurement. First he holds his hand, with palm to rear and fingers pointing upward, at such distance from his eye that each finger covers the 1-foot space between two vertical lines. Then he lowers his hand to his side without changing the angle of the wrist or elbow and notes the exact point at which the hand strikes the body. Thereafter when measuring with the fingers he first places his hand at this point and raises his arm to the front without changing the angle of the wrist or elbow. His hand will then be in the correct position for measuring "fingers." The men then determine the proper distance of fingers from the eye as explained by the instructor.

(c) The instructor next explains and demonstrates the use of sights in lateral measurement. The procedure parallels that described in *b* above for measuring by fingers. The instructor explains that the sight leaf is raised and the rifle pointed at the vertical lines on the wall. The eye is then moved along the stock of the rifle until the point is found where the raised sight leaf exactly covers the 1-foot space between two of the vertical lines. The men then take position on the testing line and each determines the proper distance of his eye from the sight and marks this position on the stock of his rifle or carefully notes it. For the average man the position of the eye is about 14 inches from the sight.

(*d*) Practice in lateral measurement using fingers and sights is given.

c. No. 3.—(1) *Purpose.*—To afford practice in target designation by pointing with the rifle.

(2) *Method.*—(*a*) The squad is formed faced to the rear. The instructor then points out the target to the squad leader, who takes the kneeling or prone position, estimates the range, adjusts his sight, alines his sights on the target, and then calls "Ready."

(*b*) The members of the squad then move in turn to a position directly behind the squad leader and look through the sights until they have located the target. The range is given orally by the squad leader to each individual.

(*c*) As soon as each man has located the target he moves to the right or left of the squad leader, sets his sight, places his rifle on a bayonet rest, and alines his sights on the target.

(*d*) The instructor, assisted by the squad leader, verifies the sight setting and the alinement of the sights of each rifle.

d. No. 4.—(1) *Purpose.*—To afford practice in target designation by oral description.

(2) *Method.*—(*a*) The squad is deployed faced to the rear. The squad leader is at the firing point, where bayonet rests have been provided for each rifle.

(*b*) At a prearranged signal the target is indicated by the display of a flag. When the squad leader states that he understands the position of the target the flag is withdrawn.

(*c*) The squad is then brought to the firing point, placed in the prone position, and each man required to set his sight, use the bayonet rest, and sight his rifle on the target according to the oral description of the squad leader. The squad leader gives his target designation from the prone position.

(*d*) The squad leader's designation is checked from the ground. The men are required to leave their rifles on the rests, properly pointed, until checked by the instructor or squad leader.

SECTION IV

RIFLE FIRE AND ITS EFFECT

■ 117. TRAJECTORY.—*a. Nature.*—The trajectory is the path followed by a bullet in its flight through the air. The bullet leaves the rifle at a speed of 2,700 feet per second. Because of this great speed the trajectory at short range is almost straight or flat.

b. Danger space.—The space between the rifle and the target in which the trajectory does not rise above a man of average height is called the danger space. The trajectory for a range of 700 yards does not rise above 68 inches. Therefore, it is said that the danger space for that range is continuous between the muzzle of the gun and the target. For ranges greater than 700 yards, the bullet rises above the height of a man standing, so that only parts of the space between the gun and the target are danger spaces. (Fig. 52.)

■ 118. DISPERSION.—Because of differences in ammunition, aiming, holding, and wind effects, a number of bullets fired from a rifle at a target are subject to slight dispersion. The trajectories of those bullets form an imaginary cone-shaped figure called the cone of dispersion.

■ 119. SHOT GROUPS.—When the cone of dispersion strikes a vertical target it forms a pattern called a vertical shot group. A shot group formed on a horizontal target is called a horizontal shot group. Owing to the flatness of the trajectory, horizontal shot groups on level ground vary in length from 100 to 400 yards depending upon the range.

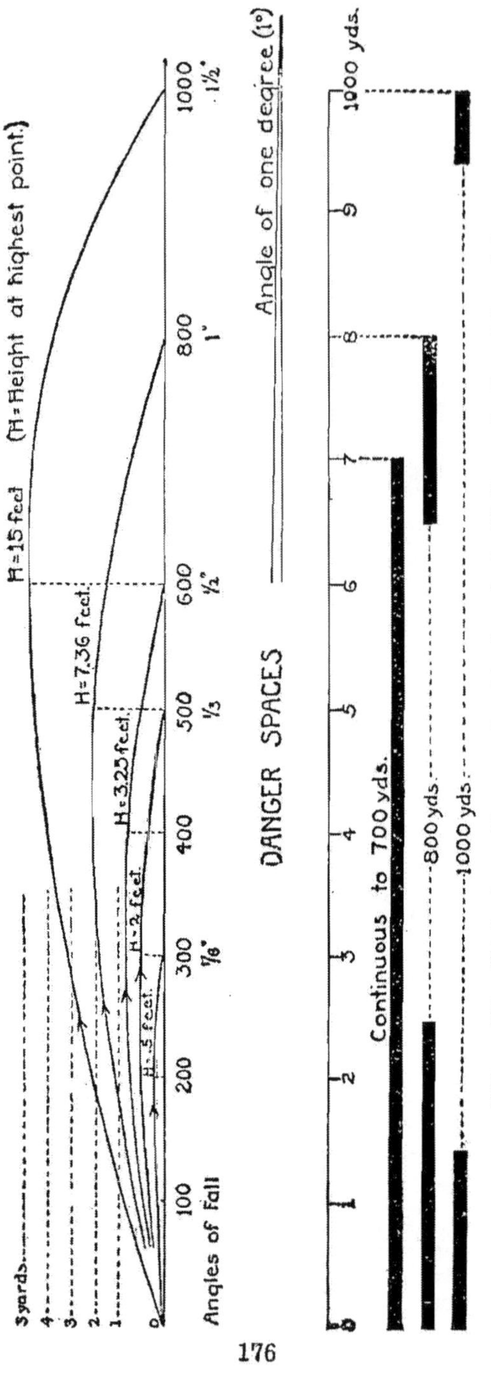

FIGURE 52.—Trajectory diagram (vertical scale is 20 times horizontal scale).

■ 120. BEATEN ZONE.—The beaten zone is the area on the ground struck by the bullets forming a cone of dispersion. When the ground is level, the beaten zone is also a horizontal shot group. The slope of the ground has great effect on the shape and size of the beaten zone. Rising ground shortens the beaten zone. Ground that slopes downward and in the approximate curve of the trajectories greatly lengthens the beaten zone. Falling ground with greater slope than the trajectory will escape fire and is said to be in defilade.

■ 121. CLASSES OF FIRE.—*a*. Fire is classified with reference to its direction as—

(1) *Frontal.*—Fire delivered on the enemy from his front.

(2) *Flanking.*—Fire delivered on the enemy from his flank.

b. Fire is classified with reference to its trajectory as—

(1) *Grazing.*—Fire approximately parallel to the ground and ordinarily not more than the average height of a man (68 inches) above it.

(2) *Plunging.*—Fire in which the angle of fall of the bullets with reference to the slope of the ground is such that the danger space is practically confined to the beaten zone and the length of the beaten zone is materially lessened. Fires delivered from high ground on ground lying approximately at right angles to the cone of fire, or against ground rising abruptly to the front with respect to the position of the rifle, are examples of plunging fire. As the range increases, fire becomes increasingly plunging because the angle of fall of the bullets becomes greater.

(3) *Overhead.*—Fire delivered over the heads of friendly troops.

c. Flanking fire is more effective than frontal fire. Grazing fire is more effective than plunging fire because the beaten zone is much longer. Overhead fire with the rifle is unusual; it should only be used when the ground affords protection to the friendly troops.

■ 122. EFFECT OF FIRE.—*a*. By making use of cover and the fires of supporting weapons, rifle units work as close to the enemy as possible before opening fire. Rifle fire should not be opened at ranges greater than 600 yards and only then if rifle fire must be resorted to in order to continue the advance.

b. A ricochet is effective if it strikes a man soon after it rises from the ground.

c. Rifle fire is effective against low-flying airplanes. The effect of fire on moving targets is covered in chapters 3 and 4.

d. Even though hits can no longer be made, fire is often continued against the enemy's position to keep him under cover and reduce the effectiveness of his fire.

e. When opposing forces are entrenched and neither side is trying to advance, fire for moral effect alone is of no value.

■ 123. EXERCISE.—*a. Purpose.*—To show trajectories.

b. Method.—The unit under instruction watches the firing of a few tracer bullets at targets at 300, 600, and 800 yards. The flatness of the trajectory is pointed out.

SECTION V

APPLICATION OF FIRE

■ 124. GENERAL.—Fire and movement are combined in the combat action of rifle units. The effective application of fire by these units is essential to their success.

a. Application of fire in attack.—In the attack the fire of a rifle unit is generally advanced by small groups or by individuals. Irregular formations are used. The unit applies its fire progressively by individuals and small groups that work their way forward to new firing positions covered by the fire of men in place. The squad and smaller groups must be trained to place a large volume of accurate fire upon probable enemy locations and indistinct or concealed targets such as enemy machine guns and small groups. Squads and smaller groups must be trained to apply their fire quickly and accurately upon order or signal of their leaders and in appropriate circumstances without orders.

b. Application of fire in defense.—In the defense, fire is delivered by small groups and individuals from positions which they must hold. The disposition of these small groups in platoon areas must be adapted to the terrain, be inconspicuous and hard to see. In addition they must have good fields of fire and must take full advantage of available cover and concealment.

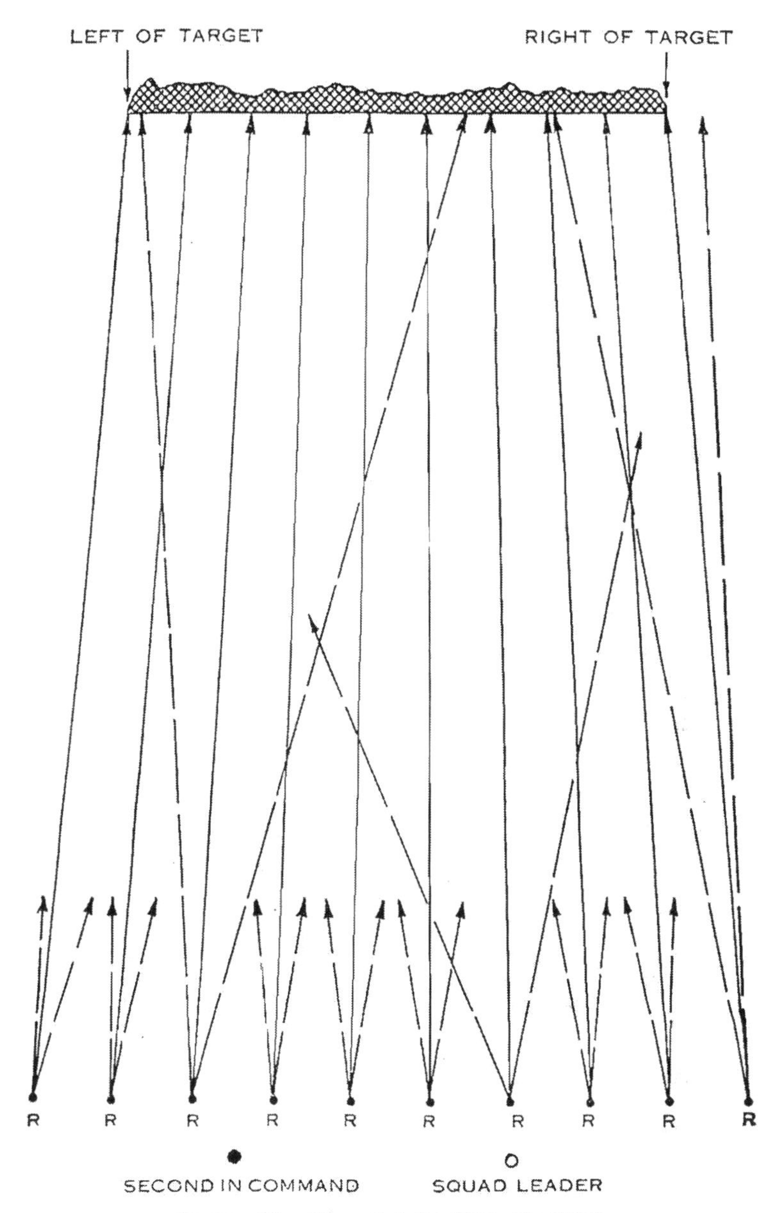

FIGURE 53.—Fire distribution of squad.

NOTE.—Each "R" in figure represents a rifleman of the squad in position. Solid lines indicate direction of fire for first shot; broken lines indicate sectors of fire.

■ 125. CONCENTRATED AND DISTRIBUTED FIRE.—*a. Concentrated fire.*—Concentrated fire is fire directed at a single point. This fire has great effect but only at a single point. Machine guns and other automatic weapons are examples of suitable targets for concentrated fire.

b. Distributed fire.—(1) Distributed fire is fire distributed in width for the purpose of keeping all parts of a target under fire. It is habitually used on targets having any considerable width.

(2) When a squad distributes its fire each rifleman fires his first shot on that portion of the target corresponding generally to his position in the squad. He then distributes his remaining shots to the right and left of his first shot, covering that part of the target on which he can deliver accurate fire without changing his position. The amount of target each rifleman can cover depends upon his position and the range. In some situations each rifleman can cover the entire target. Fire is not limited to points known to contain an enemy; on the contrary, riflemen space their shots so that no portion of the target remains unmolested. This method of fire distribution is employed without command. It enables squad leaders to distribute the fire of their units so as to keep the entire target under fire. (See fig. 53.)

(3) If a squad is employing this method of fire distribution and other targets appear, the squad leader announces such changes in the fire distribution as are necessary.

(4) If a whole platoon is engaging the same target, all squads of the platoon distribute their fire in the same manner.

■ 126. ASSAULT FIRE.—Assault fire is that fire delivered by a unit during its assault on a hostile position. Riflemen with bayonets fixed and taking full advantage of existing cover, such as tanks, boulders, trees, walls, and mounds, advance rapidly toward the enemy and, as they advance, fire at areas known or believed to be occupied by hostile personnel. This fire is usually delivered from the standing position and is executed at a rapid rate.

■ 127. RATE OF FIRE.—*a.* The soldier fires at the rate of fire most effective under existing conditions. To exceed this rate is to waste ammunition.

b. In rapid fire practice the soldier is trained to shoot from 7 to 10 shots per minute, depending upon the range. This rate is increased to 10 to 15 shots per minute as the soldier gains experience.

■ 128. FIRE DISCIPLINE.—*a*. Fire discipline is a state of order, coolness, efficiency, and obedience existing among troops engaged in a fire fight. When fire discipline is good, men fight as they have been trained to fight and obey orders promptly and carefully; they resist and overcome the influence of danger, excitement, and confusion. Fire discipline is necessary for proper control by leaders and upon this control depend the teamwork and effectiveness of the collective fire of the unit. The training necessary to insure good fire discipline cannot be completed during the brief period devoted to technique of fire. Training in fire discipline starts with the soldier's first drill and continues throughout his military training. Any drill or exercise which develops alertness and the habit of obedience or other soldierly qualities will aid in developing the character essential to fire discipline.

b. Fire discipline is maintained by leaders chiefly by their example of coolness and courage. Replacement of casualties is an element of fire discipline which keeps the unit working as a team in spite of losses. If any group finds itself without a leader, it is essential that one of the group assume leadership and carry out the mission or attach the group to the nearest organized unit. An individual separated from his squad fights on his own initiative only when he has reason to believe that his single effort will accomplish some important result. Otherwise he reports to the nearest leader at once.

c. Fire discipline in the squad is the responsibility of the squad leader; he is assisted by the second-in-command. During the fire fight the squad leader posts himself where he can best control his squad. The second-in-command posts himself where he can best assist the squad leader.

■ 129. FIRE CONTROL.—*a*. Fire control pertains to the squad or smaller groups. Its application to a rifle unit as large as a platoon is exceptional. Fire control consists of the initiation and supervision of the fire of the squad or a smaller group by its leader. By initiating such fire on order or signal the full effect of surprise can be secured. On the other hand the irregular formations adopted for an advance often make it necessary for fire to be opened and maintained on the initiative of individuals as circumstances require. However, the leader of the squad or smaller group must supervise and seek to control the fire of his men so that their fire is directed and maintained at suitable targets. All must understand that controlled fire is always the most effective.

b. Squad leaders, assisted by their seconds-in-command, exercise fire control by means of orders and signals. The signals most frequently used are—

Signals for range.
Commence firing.
Fire faster.
Fire slower.
Cease firing.
Are you ready?
I am ready.
Fix bayonets.

■ 130. FIRE ORDERS.—*a. Definition.*—The instructions by which the fire of a squad is directed and controlled form the fire order.

b. Basic elements of fire order.—A fire order contains three basic elements which are announced or implied. Only those elements or those parts of them that are essential are included. The sequence is always—

Target-designation element.
Fire-distribution element.
Fire-control element.

(1) *Target-designation element.*—The target is designated as described in section III.

(2) *Fire-distribution element.*—(*a*) The fire-distribution element is usually omitted from the fire order. The method of fire distribution described in paragraph 125*b* is used in the absence of instructions to the contrary. When necessary, the fire-distribution element includes the subdivision of the target.

(*b*) For example, a squad leader desires to engage two machine-gun nests; the fire-distribution element of his order is indicated by the italicized words below:

Range: 500.
Front.
Machine gun at base of lone pine.
Cooper, Emerson, Crane, Hines, Jones, your target.
Range: 500.
Left flank.
Machine gun at base of haystack.
Brown, Smith, Turner, Howard, Stone, your target.

(*c*) The squad leader may engage two targets by placing half of the squad under the command of the assistant squad leader and directing him to engage one target, while he engages the other target with the other half of the squad.

(3) *Fire-control element.*—Initially the fire-control element may consist only of the command or signal COMMENCE FIRING or FIRE AT WILL. It may include the number of rounds. For example—

AT MY SIGNAL (followed by hand signal).
FIVE ROUNDS, FIRE AT WILL.

(4) *Example of a complete fire order.*

(*a*) *Target-designation element.*

Range	Range: 500.
Direction	Reference: right edge of lone building. Right one finger.
Description of target	Target: group of enemy.

(*b*) *Fire-distribution element.*—Implied.

(*c*) *Fire-control element.*—FIRE AT WILL.

■ 131. DUTIES OF LEADERS.—The following summary of duties of leaders relates only to their duties in the technique of fire.

a. Squad leader.—(1) Carries out orders of platoon leader.

(2) Selects firing positions for squad.

(3) Designates targets and issues fire orders.

(4) Controls fire of squad.

(5) Maintains fire discipline.

(6) Observes targets and effect of fire.

b. Second-in-command.—(1) Carries out orders of squad leader.

(2) Helps the squad leader maintain fire discipline.

(3) Assumes command of squad in absence of squad leader.

(4) Participates in firing when the fire of his rifle is considered more important than other assistance to the squad leader.

SECTION VI

LANDSCAPE-TARGET FIRING

■ 132. SCOPE AND IMPORTANCE.—*a.* After satisfactory progress has been made in the preceding steps, the soldier may practice what he has learned by firing at landscape targets.

b. The advantages of landscape-target firing are—

(1) It permits close supervision of all members of the firing unit.

(2) It clearly and quickly demonstrates the application and effect of fire.

(3) It can be conducted indoors when lack of facilities or weather conditions make this desirable.

c. In circumstances where there is a choice between landscape-target firing and field-target firing, the latter will be selected. Therefore, landscape-target firing is not required as a part of training.

■ 133. DESCRIPTION OF LANDSCAPE-TARGET.—A landscape-target is a panoramic picture of a landscape. It is so drawn that all or nearly all of the salient features are recognizable at a distance of 1,000 inches. The standard target is the series A target of five sheets in black and white.

■ 134. WEAPONS TO BE USED.—Firing at landscape targets should be conducted with caliber .22 rifles, preferably the M1922M2 equipped with the Lyman receiver sight. If there are not enough caliber .22 rifles available, caliber .30 rifles may be used.

■ 135. PREPARATION OF TARGETS.—*a. Mounting.*—(1) The sheets are mounted on frames made of 1- by 2-inch dressed lumber, with knee braces at the corners. The frames are

24 by 60 inches and are covered with target cloth which is tacked to the edges.

(2) The target sheets are mounted as follows: Apply a thin coat of flour paste to the target cloth and let it dry for about an hour; apply a thin coat of paste to the back of the paper sheet and let it dry about an hour; apply a second coat of paste to the back of the paper and mount it on the cloth; smooth out wrinkles, with a wet brush or sponge, working from the center to the edges. The frame must be placed on some surface that will keep the cloth from sagging when the paper is pressed on it. A form for this purpose can be easily constructed. It must be of the same thickness as the lumber from which the frames are built and must have approximately the same dimensions as the aperture of the target frame.

b. Target frames.—Panels (mounted as described above) are set in a vertical frame made of 4- by 4-inch posts sunk upright in the ground and spaced 5 feet from center to center. Horizontal 2 by 4's add stability, and form a base to support the panels. Cleats and dowels hold the panels and permit easy removal.

c. Range indicators.—To permit proper designation of targets, assumed ranges must be used on the landscape target. Small cards with appropriate ranges are tacked along one or both edges of a series of panels. The firers are warned that ranges are announced for the sole purpose of designating the target, and that the zero sight setting of their rifles for the 1,000-inch range must not be changed.

d. Direction cards.—In order to provide the direction element in oral target designation, small cards on which are painted "Front," "Right front," "Left front," "Right flank," "Left flank," are tacked above the appropriate panels of the landscape series.

e. Scoring devices.—(1) Scoring the exercises tends to create competition between squads and enables the instructor to grade their relative proficiency. A scoring device conforming in size to the 50 and 75 percent shot groups to be expected of average shots firing at 1,000 inches can be easily made from wire; or a better one can be made by imprinting a scoring diagram on a sheet of transparent celluloid. The scoring space is outlined on the target in pencil before the target is shown to squad leaders. This procedure prevents squad leaders from misunderstanding the limits of the designated target. Upon completion of firing, the entire squad is shown the target and the results of the firing.

(2) Although shot groups take the form of a vertical ellipse, the 50 and 75 percent zones should be shown by the devices as rectangles. This is for convenience in their preparation. At 1,000 inches the 50 percent zone is a rectangle 2.5 inches high by 2 inches wide and the 75 percent zone is a rectangle 5 inches high by 4 inches wide. At 50 feet the 50 percent zone is 1.5 inches high by 1.2 inches wide and the 75 percent zone is 3 inches high by 2.4 inches wide. The target is at the center of the inner rectangle or 50 percent zone.

(3) For a linear target over which the automatic rifleman is to distribute his fire, the 50 percent zone is formed by two parallel lines, drawn parallel to the longer axis of the target with the target midway between those lines. At 1,000 inches the lines are 2.5 inches apart; at 50 feet they are 1.5 inches apart. Two additional lines similarly drawn form the 75 percent zone. At 1,000 inches the lines are 5 inches apart; at 50 feet they are 3 inches apart. The width of the zones vary according to the size of the target selected. At 1,000 inches the zones extend 1 inch beyond each end of the target; at 50 feet they extend 0.6 inch beyond each end of the target. The zones are divided into a convenient number of equal parts. The number depends on the length (width) of the target and the number of men firing. This is done in order to give a score for distribution of shots fired on a linear target (see par. 138*b*).

■ 136. ZEROING-IN OF RIFLES.—*a*. It is necessary to zero-in those rifles that are to be used in landscape target firing. A blank target with a row of ten 1-inch-square black pasters about 6 inches from and parallel to the bottom edge of the target should be used for this purpose. In all firing for zeroing-in, sandbag rests are used.

b. The procedure is as follows:

(1) Sights of all rifles are blackened.

(2) The squad is deployed on the firing point; the squad leader takes his position in rear of the squad.

(3) The instructor causes each firer to set his sights at zero elevation, or 100 yards, and checks each rifle.

(4) Each man is assigned as an aiming point that paster which corresponds to his position in the squad.

(5) At the command of the instructor, rifles are loaded with three rounds.

(6) At the command THREE ROUNDS, FIRE AT WILL, each man fires three shots at his spotter.

(7) The instructor then commands: CEASE FIRING, OPEN BOLTS. The squad leader checks to see that this is done.

(8) The instructor and squad leader inspect the target and, based upon the location of the center of impact of the resultant shot group, give each man the necessary sight correction. For example, with the caliber .22 rifle, "Up 1 minute, right ½ point"; or with the caliber .30 M1917 rifle, "Up 100, left one inch."

(9) The firing continues as outlined above until all rifles are zeroed-in, that is, until each man has hit his aiming paster.

c. For the caliber .22 rifle with the Lyman receiver sight, a change of 5 minutes in elevation moves the strike of the bullet about 1½ inches at a range of 1,000 inches. A change of one point of windage moves the strike about 1¼ inches. At 50 feet a change of 6 minutes in elevation moves the strike of the bullet about 1 inch, and a change of one point of windage, about ¾ inch. For the caliber .30 rifle, at a distance of 1,000 inches, a change of 100 yards in elevation moves the strike of the bullet about 4 inches. At 50 feet a change of 100 yards in elevation moves the strike of the bullet about 2⅖ inches.

■ 137. FIRING PROCEDURE.—The following sequence is used in conducting firing exercises:

a. All members of the squad, except the squad leader, face to the rear.

b. The instructor takes the squad leader to the panels and points out the target to him.

c. They return to the firing point; the squad leader takes charge of his squad and directs his men to resume their firing positions.

d. The squad leader gives the command: LOAD, cautioning "______ rounds per rifleman only."

e. The squad leader designates the target orally. Reference to panels to indicate direction is not allowed. To complete the fire order, the squad leader adds: FIRE AT WILL.

f. When the squad has completed firing, the squad leader commands: CEASE FIRING, OPEN BOLTS. The squad then examines the target. The target panel is removed, scored, marked with the squad number, and replaced with a new panel.

g. The instructor holds a short critique after each exercise.

■ 138. SCORING.—*a*. *Concentrated fire*.—In concentrated fire the sum of the value of the hits within the two zones is the score for the exercise. For convenience of scoring and comparison, 100 is fixed as the maximum score. Any method of scoring and of distribution of ammunition among members of the squad may be used. For example—

(1) Number of rounds fired, 50.

(2) Value of each hit in 50 percent zone, 2.

(3) Value of each hit in 75 percent zone, 1.

b. *Distributed fire*.—A suggested method of scoring for distributed fire follows:

(1) Number of rounds fired, 50.

(2) Value of each hit in 50 percent zone, 2.

(3) Value of each hit in 75 percent zone, 1.

(4) Value of each distribution space (if target is divided into 10 equal spaces), 10.

(5) The score for distribution, plus the value of all hits, divided by two is the score for the exercise.

■ 139. EXERCISES.—*a*. *No. 1*.—(1) *Purpose*.—To teach target designation and to show the effect of concentrated fire.

(2) *Method*.—The squad leader directs the fire of his squad at a point-target indicated to him by the instructor.

b. No. 2.—(1) *Purpose.*—To teach target designation and the division of the squad fire on three points of concentration.

(2) *Method.*—The instructor indicates three point-targets to the squad leader, giving the nature of each. The squad leader applies the fire of his squad on the three targets in the proportions directed by the instructor. The score is computed for each of the three targets and these are then combined to obtain the score for the exercise.

c. No. 3.—(1) *Purpose.*—To teach target designation and fire control by switching part of the fire of the squad to a suddenly appearing target.

(2) *Method.*—The instructor indicates a point-target to the squad leader. After firing has commenced, the instructor indicates and gives the nature of a new target to a flank. The squad leader applies the fire of his squad to the first target. When the second target is indicated, he shifts the fire of the number of riflemen, as directed by the instructor, from the first to the second target.

d. No. 4.—(1) *Purpose.*—To teach the application of fire on an enemy group marching in formation, the fire control necessary to obtain fire for surprise effect, and to show the effect of fire on troops in formation.

(2) *Method.*—The instructor indicates to the squad leader a target that represents a small group of the enemy marching in approach march formation, patrol formation, or the like, the enemy not being aware of the presence of the squad. The squad leader applies the fire of his squad; his instructions must result in the simultaneous opening of fire of all weapons and the distribution of fire over the entire target. The assignment of the five riflemen on the right to fire at the rear half of the target, and the remaining riflemen at the forward half, is a satisfactory method of distributing fire over such target.

e. Training of second-in-command.—The second-in-command of the squad is given instruction and practice in similar exercises.

Section VII

FIELD-TARGET FIRING

■ 140. Scope.—The training in this phase is similar to that given the soldier in landscape-target firing, but with the added features of the use of cover, range estimation, firing the rifle with ball ammunition at field targets at unknown ranges, and fire control under more difficult conditions. Training must be progressive. The soldier is first afforded an opportunity to fire at exposed or partially concealed targets and later at targets which are completely concealed from view but exposed to fire. This training should be conducted by squads or smaller groups.

■ 141. Terrain.—*a.* Where possible, varied ground suitable for the employment of all weapons of the rifle unit is selected. It is a great advantage from the instructional standpoint to use ground that is unfamiliar to the unit to be trained.

b. In the absence of other facilities, the known-distance ranges can be used by arranging the exercises so that they will begin off the range and require the delivery of fire on the range and in a safe direction.

■ 142. Targets.—*a.* Targets may be improvised from available material or they may be obtained from the Ordnance Department.

b. With the field targets furnished by the Ordnance Department a stationary target may be represented by E or F targets placed on staves and driven in the ground.

c. A surprise target that can appear and disappear may be made by using either E or F targets fastened to an I-beam and operated by a man in a pit.

d. A movable field target may be made by fastening E or F targets to a sled (see fig. 54).

e. In the field, targets should be placed in locations that would be used by an intelligent enemy. They should not be prominently exposed or in a regular line. The exposure of targets kept out of sight at the beginning of an exercise may be indicated by the firing of blank ammunition or the operation of other noise or smoke-producing equipment in the vicinity of the target when it does appear. In platoon problems, targets may be placed so that they are visible with field glasses but invisible to the naked eye, thus requiring skill in designating the target and adjusting the fire.

f. The appearance of the targets from the firing line depends a great deal upon the direction of the sun, the background, and the angle at which the targets are placed. These factors should be taken into consideration when placing targets for an exercise.

■ 143. SHELTER.—Ranges for combat firing exercises can be efficiently operated without an elaborate system of shelters and dugouts. Simple pits to accommodate the target operators are sufficient. Every effort should be made to avoid altering the natural appearance of the terrain when locating and constructing pits. When targets are placed in the rear of or to one side of the pits, the likelihood of ricochets falling into the pit is minimized.

■ 144. SAFETY.—*a*. In general the safety precautions used at known-distance ranges apply with equal force to instruction in firing at any field target (see AR 750–10). Safety of personnel is of primary importance in conducting exercises which require the firing of ball ammunition. To this end, exercises should be drawn to conform to the state of training of the units concerned.

b. The officer in charge of an exercise is responsible for the safety of the firing; it is his duty to initiate and enforce such precautions as he deems necessary under existing conditions. No other officer can modify his instructions without assuming the responsibility for the safety of the firing.

c. While firing, no man should be permitted to be ahead of or in rear of the firing line a distance greater than one-half the interval between himself and the man next to him. For example, if the interval between skirmishers is 10 paces, then no man should be more than five paces ahead of or behind the man next to him on either side.

d. Firing must not start until it has been ascertained that the range is clear, that pit details are not exposed, and that *all safety precautions have been complied with.* Upon completion of firing, the officer in charge causes all rifles and

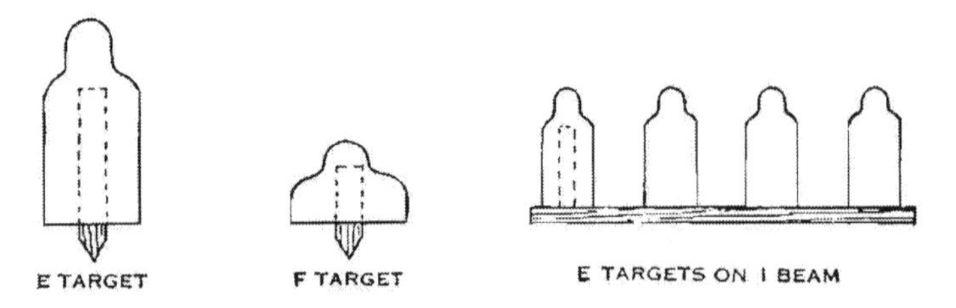

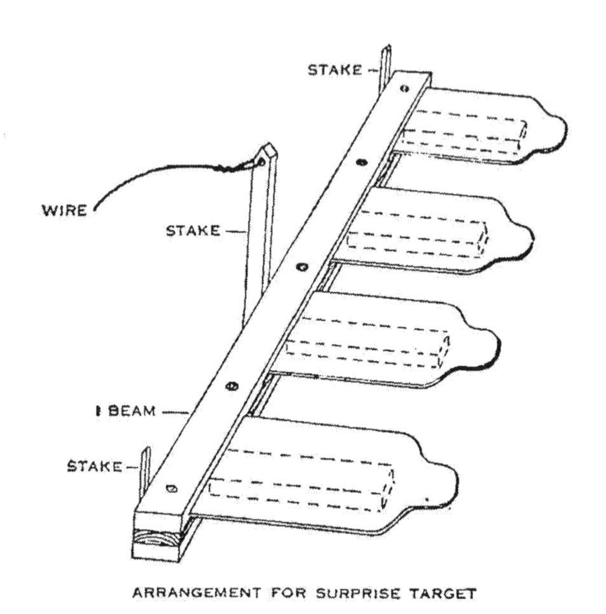

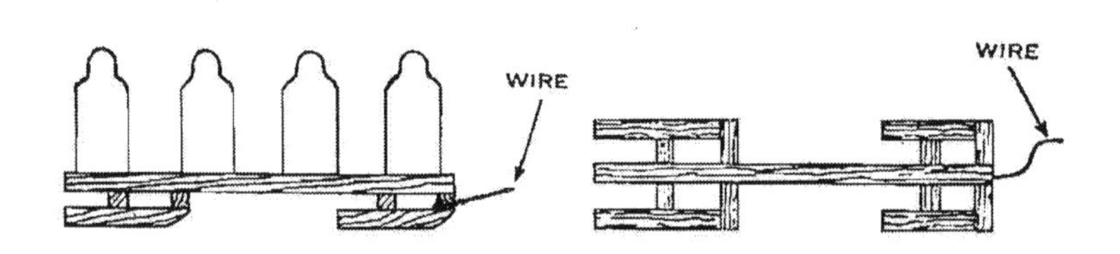

FIGURE 54.—Field targets.

belts to be unloaded and inspected, and all ammunition collected so that none remains in the possession of the men returning to camp or barracks.

e. During the firing of exercises, rifles are pointed in the direction of the target at all times. Special vigilance is required to enforce this rule when men are using cleaning rods to remove obstructions from chambers.

■ 145. GENERAL CONSIDERATIONS.—*a. Progressive training.*—The inclusion of training in moving from an approach march formation or place of concealment to firing positions is primarily to teach the soldier the proper use of cover and selection of firing positions, and to combine the technique of applying and controlling collective fire with scouting and patrolling and other allied subjects.

b. Firing positions and representation of enemy.—In battle, a unit is not deployed with individuals abreast and at regular intervals. The selection of individual and group positions is governed by the field of fire, cover or concealment while firing, covered approaches to those positions, fire control, and nature of target. Targets representing the enemy should be disposed to conform to these practices.

c. Use of cover.—(1) The proper use of cover is important for two reasons. The man who neglects the use of cover will be seen and hit. His squad not only has its fire power decreased by one rifle, but its position is unnecessarily disclosed and other casualties may follow.

(2) The use of cover and concealment is taught in scouting and patrolling. In training in firing at field targets the principles are the same.

(3) In seeking cover in a firing position men may move a few yards in any direction, but they must not be allowed to bunch behind concealment which does not afford protection from fire. They avoid positions which will mask the fire of others or cause their own fire to be dangerous to other men of their unit.

d. Marksmanship applied.—(1) The principles of rifle marksmanship are followed in this training so far as they fit the conditions.

(2) These principles should be applied to the technique of fire and to combat in a common-sense way. It should

be appreciated that the conditions encountered in combat differ from those on the target range. On the target range the soldier is not allowed to rest his rifle against an unauthorized object while firing. But in field-target firing and in battle the soldier takes advantage of trees, rocks, or any other rest which will make his fire more accurate. The positions prescribed in rifle marksmanship are used whenever the ground permits, but on rough ground it is often necessary to modify them in order to get a comfortable and steady position. The loop sling is preferable except—

(*a*) In the standing position.

(*b*) When the situation requires readiness for immediate use of the bayonet.

(*c*) In emergencies demanding immediate fire without time for adjustment of the loop sling.

e. Use of battle peep sight.—The battle peep sight corresponds to a sight setting of 400 yards. It is used only when time is lacking for setting the peep sight. (For firing at aircraft or other moving targets (see pars. 67*b* and 81*c*.) By keeping the peep sight always set at 500 yards when not in use, the soldier has two sights ready for emergencies.

SECTION VIII

FIRE EXERCISES

■ 146. GENERAL.—*a*. Exercises for field-target firing should conform to the terrain used. One or more problems should be fired while wearing the gas mask.

b. Problems should start with the unit—

(1) Already deployed in a firing position, or

(2) Halted in an approach march formation (or in a place of concealment) with scouts present in formation, or

(3) Advancing in an approach march formation with scouts out.

c. (1) If the exercise begins with the unit already deployed in a firing position, each man should occupy a selected position with special attention to cover and concealment.

(2) If the exercise begins with the unit halted in an approach march formation, squad leaders lead their squads forward by covered routes and send their riflemen to firing

positions by individual directions. Occupation of the initial firing position is executed with the minimum of exposure.

(3) If the exercise begins with the unit advancing in an approach march formation, the scouts must be checked by assumed enemy fire when they are at a suitable location for a firing position of the squad. If the scouts are permitted to advance beyond this position, they must be withdrawn from the exercise before the squad opens fire.

■ 147. CRITIQUE.—At the completion of each exercise the instructor conducts a critique. A suggested form for this critique follows:

a. Purpose of the exercise.

b. Orders of squad leader.

c. Approach and occupation of the firing position (individual concealment and cover).

d. Action of individuals.

e. Rate of fire.

f. Fire control.

g. Effect of the fire (upon completion of firing, the range being clear, the targets are scored).

h. Performance of the unit—satisfactory or unsatisfactory.

■ 148. SUGGESTED EXERCISES.—*a*. *No. 1*.—(1) *Purpose*.—Practice in fire orders, application of the fire of a squad in position, fire control, proper individual concealment in the occupation of the firing position.

(2) *Method*.—Enemy is represented by one group of targets exposed to fire but partially concealed from view, requiring a simple fire order. Squad leader is shown the targets (personnel with flag) and safety limits for firing position of the squad. When the squad leader fully understands the location and nature of the target and when the instructor informs him that the range is clear, he loads ball ammunition, gives the fire order, and fires the problem. The range should be estimated by eye and the target designated by oral description.

b. *No. 2*.—(1) *Purpose*.—Practice in fire orders, application of the fire of a rifle squad on a linear target, fire control, proper deployment and individual concealment in the occupation of the firing position, engagement of a surprise target.

(2) *Method*.—Silhouette targets, representing an enemy squad deployed in a firing position, are partially concealed

from view but exposed to fire. A screen behind the targets is marked with distribution spaces to give the squad credit for those shots that did not hit the targets but which would have had an effect on an enemy. The squad is in rear of the firing position. The squad leader (scouts) is shown the linear target (by flag). He then leads his squad forward and disposes it in a concealed firing position. When the squad leader is told the range is clear he engages the target with surprise fire. A surprise target, well to the flank of the first target, representing an enemy machine gun, appears shortly after the squad has engaged the linear target. The squad leader is told the amount of fire to shift to the surprise target. In the critique, the matters suggested in paragraph 147 should be discussed and also the proper distribution of the fire of a rifle squad on a linear target and the engagement of the surprise target.

c. No. 3.—(1) *Purpose.*—Practice in target designation by scouts with tracer ammunition, and squad instruction in approaching and assuming a firing position.

(2) *Method.*—The squad is marching in approach march formation with both scouts well in advance. When the scouts reach the firing position they observe the targets representing an enemy group about 400 yards to their front. They determine the range by firing on the target with tracer bullets. The squad leader conducts his squad forward, establishes the men in firing positions, and engages the targets with the proper class of fire after the scouts have designated the target with tracers. Special attention is paid to the use of cover and concealment by all men while moving up and during the selection and occupation of positions.

d. No. 4.—(1) *Purpose.*—Practice in firing at moving targets.

(2) *Method.*—Riflemen fire individually at targets carried on long sticks by men in the pits of a class A range. The men in the pits are each assigned a space (about the width of five regular range-target spaces) in which they walk back and forth continuously. By whistle signal, targets are exposed to the firing line for 5 seconds and then concealed for 5 seconds. Targets are exposed once for each shot to be fired. On the firing line one man is assigned to each target. Ranges of 200 or 300 yards are best suited for this exercise.

CHAPTER 6

ADVICE TO INSTRUCTORS

Paragraphs

SECTION I

GENERAL

■ 149. PURPOSE.—The provisions of this chapter are to be accepted as a guide and not as having the force of regulations. They are particularly applicable to emergency conditions when large bodies of troops are being trained under officers and noncommissioned officers who are not thoroughly familiar with approved training methods.

SECTION II

MECHANICAL TRAINING

■ 150. CONDUCT OF TRAINING.—*a*. Instruction is so conducted as to insure the uniform progress of the unit.

b. The instructor briefly explains the subject to be taken up and demonstrates it himself or with a trained assistant.

c. The instructor then has one man in each squad or subgroup perform the step while he explains it again.

d. The instructor next requires all members of the squads or subgroups to perform the step under the supervision of their noncommissioned officers. This is continued until all men are proficient in the particular operation, or until those whose progress is slow have been placed under special instructors.

e. Subsequent steps are taken up in like manner during the instruction period.

SECTION III

MARKSMANSHIP—KNOWN-DISTANCE TARGETS

■ 151. GENERAL.—Training is preferably organized and conducted as outlined in paragraphs 41 and 42. Officers should be considered as the instructors of their units. Since only one step is taken up at a time, and since each step begins with a lecture and a demonstration showing exactly what to do, the trainees, although not previously instructed, can carry on the work under the supervision of the instructor.

■ 152. PLACE OF ASSEMBLY FOR LECTURES.—Any small ravine or cup-shaped area makes a good amphitheater for giving the preliminary lectures if no suitable building is available.

■ 153. ASSISTANT INSTRUCTORS.—*a.* It is advantageous to have all officers and as many noncommissioned officers as possible trained in advance in the prescribed methods of instruction. When units are undergoing marksmanship training for the first time, this is not always practicable nor is it absolutely necessary. A good instructor can give a clear idea of how to carry on the work in his lecture and demonstration preceding each step. In the supervision of the work following the demonstration, he can correct any mistaken ideas or misinterpretations.

b. When an officer in charge of rifle instruction is conducting successive organizations through target practice, it is advisable to attach to the first organization taking the course certain officers and noncommissioned officers of the organizations that are to follow. This attachment should remain in effect for the period of preparatory work and for a few days of range firing. These men act as assistant instructors when their own units take up the work. They are particularly useful when one group is firing on the range and another is going through the preparatory exercises, both under the supervision of one instructor.

■ 154. EQUIPMENT.—The instructor should personally inspect the equipment for the preparatory exercises before the training begins. A set of model equipment should be prepared in advance by the instructor for the information and guidance of the organization about to take up the preparatory work. The sighting bars must be made as described, and the hole representing the peep sight must be absolutely circular. If the sights are made of tin, the holes should be bored by a drill. Good rear sights can be made for the sighting bars by using cardboard and cutting the holes with a punch for cutting wads for 10-gage shotgun shells. Bull's-eyes painted on a white disk are not satisfactory. Bull's-eyes cut out of black paper with a shotgun-wad cutter and pasted on white paper make satisfactory aiming points either to paste on the face of the disk or to use in position and trigger-squeeze exercises, when small gallery targets are not available for this purpose.

■ 155. INSPECTION OF RIFLES.—No man is required to fire with an unserviceable or inaccurate rifle. All rifles should be carefully inspected far enough in advance of the period of training to permit organization commanders to replace or repair those that are inaccurate or defective. Rifles that have badly pitted barrels are not accurate and should not be used.

■ 156. AMMUNITION.—The best ammunition available should be reserved for record firing and the men should have a chance to learn their sight settings with that ammunition before record practice begins. Ammunition of different makes and of different lots should not be used indiscriminately.

■ 157. ORGANIZATION OF WORK.—*a. In preparatory training.*—

(1) The field upon which the preparatory work is to be given should be selected in advance and a section of it assigned to each organization. The equipment and apparatus for the work should be on the ground and in place before the morning lecture is given, so that each organization can move to its place and begin work immediately and without confusion.

(2) Each company should be organized in two lines, facing away from each other. In this way the company officers and other instructors, whose position is normally between the lines, have all of their squads under close supervision. In figure 55 the groups represented consist of 8 men each.

(3) The arrangement of the equipment is as follows:

(*a*) On each line are placed the sighting bars and rifle rests at sufficient intervals to permit efficient work.

(*b*) Fifty feet from each line is placed a line of small boxes with blank paper tacked on one side, one box and one small sighting disk to each rifle rest.

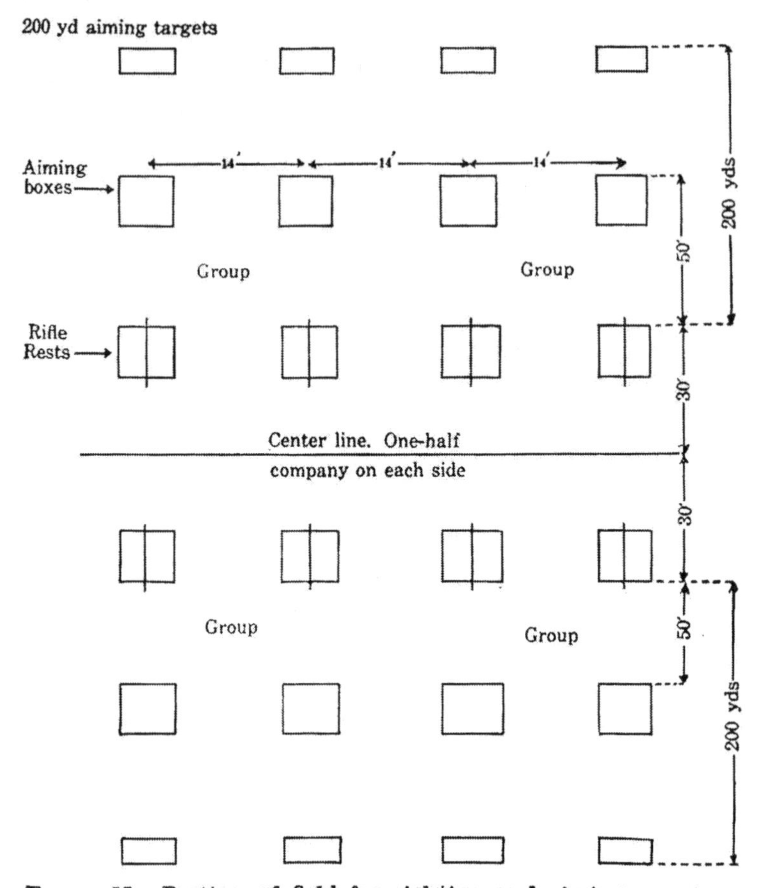

FIGURE 55.—Portion of field for sighting and aiming exercises.

(*c*) Two hundred yards from each line is placed a line of frames suitable for 200-yard shot group exercises, one frame to each group. These frames have blank paper tacked or pasted on the front. A 10-inch sighting disk is placed with each frame. Machine-gun targets make acceptable frames for this work.

(4) In position, trigger-squeeze, and rapid-fire exercise targets should be placed at 1,000 inches and 200 yards. The groups represented in figure 56 consist of 8 men each.

(5) Units will have to vary from this arrangement when sufficient level ground is not available. It will usually be found, however, that everything except the long-range shot group work can be carried on in two lines.

b. In range practice.—(1) The range work should be so organized that there is a minimum of lost time for each man. Long periods of inactivity while awaiting a turn on the firing line must be avoided. Therefore, the number of men on the range should be accommodated to the number of targets available. Six men per target is about the maximum and four the minimum for efficient handling.

(2) In both slow and rapid fire it is advisable to have the next order on the line practicing with dummy ammunition or simulated fire (see par. 56*b*(2)). When the size of the firing point makes this impracticable, each order should be given several scores of simulated fire before firing with ball cartridges.

(3) Subject to ammunition allowances, the following method of conducting range practice has been found to produce uniformly excellent results when the full allowance of time is devoted to the training:

(*a*) Organizations camp on the range, if the range is more than a mile from the post.

(*b*) Firing is begun by a group consisting of approximately half of each organization. This group is made up of those who have proved themselves the best qualified by the examination on preparatory work and those known to be good shots. The men not included in this first group make up the pit details and undergo additional preparatory training, particularly in rapid fire.

(*c*) After about 30 hours' instruction practice all of the first group, except those few who have not been shooting well, fire for record.

(*d*) The second group, made up of those who have not fired and those who were rejected from the first group, now begins firing. The men who have completed record firing perform all fatigue.

(*e*) After about 30 hours' instruction practice those of the second group who have been shooting well and who have a very good chance to qualify, fire for record.

(*f*) During the remainder of the allotted time the efforts of the officers and noncommissioned officers are concentrated on the men who were not ready to fire for record with the second group. This last group fires for record by the end of the time allotted for range practice.

(4) If the range facilities permit an entire organization to fire at one time without having more than six men per target, the same general scheme as that outlined above is followed.

(*a*) Firing is begun with all of the men of the organization taking part.

(*b*) After about 30 hours' instruction practice all except those who have not been shooting well fire for record.

(*c*) The efforts of the instructors are concentrated on the remainder of the organization for the rest of the allotted time.

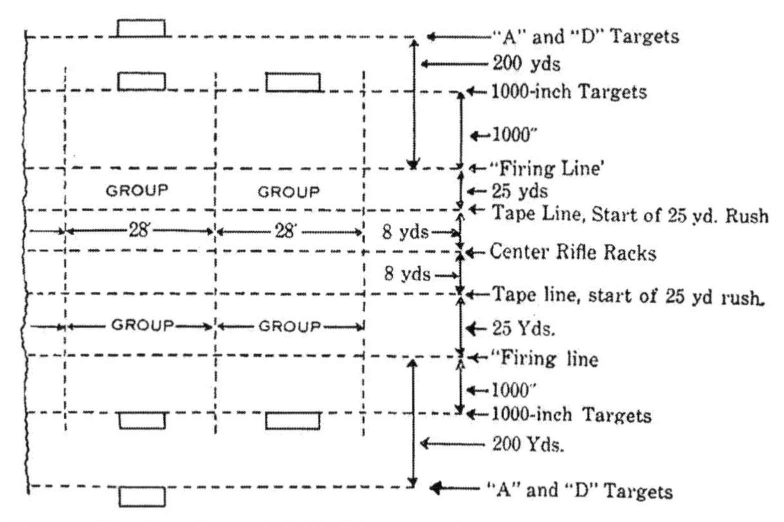

FIGURE 56.—Portion of field laid out for position, trigger-squeeze, and rapid-fire exercises.

■ 158. MODEL SCHEDULES.—*a*. *Course A*.—The following schedule of preparatory exercises is suggested as a guide for a 2 weeks' course. The practice firing is based upon a maximum of six men per target.

(1) *Lecture and demonstration.*

MONDAY

	Hours AM	Hours PM
FIRST STEP	1	
First, second, and third sighting exercises	3	
Continuation of first three exercises, including long-range shot group exercise		3
Care and cleaning of the rifle; safety precautions		1

TUESDAY

	Hours AM	Hours PM
Second step	1	
Position exercise in all positions	3	
THIRD STEP		1
Trigger-squeeze exercise using sandbag rest followed by exercises in all positions		3

WEDNESDAY

	Hours AM	Hours PM
FOURTH STEP	1	
Bolt-operation exercise	2	1
Exercises in taking position rapidly	1	
Rapid-fire exercises		3

THURSDAY

	Hours AM	Hours PM
FIFTH STEP	1	
Score-book exercise	2	
Review trigger-squeeze exercise in all positions	1	
Rapid-fire exercise		2
Final examination		2

(2) *Practice firing.*

FRIDAY

	Hours
Table I, 200 yards (simulated)	1
Table I, 200 yards	1½
Table I, 300 yards (simulated)	1
Table I, 300 yards	1½
Table I, 500 yards (simulated)	1
Table I, 500 yards	2

SATURDAY

	Hours
Tables II and III (simulated)	4

MONDAY

	Hours
Table II, 200 yards (simulated)	½
Table II, 200 yards	1
Table II, 300 yards (simulated)	½
Table II, 300 yards	1
Table II, 500 yards (simulated)	1
Table II, 500 yards	2½
Table III (simulated)	1½

TUESDAY

Table II (simulated)	1
Table II	3
Table III, 200 yards, sitting and kneeling (simulated)	½
Table III, 200 yards, sitting	1½
Table III, 200 yards, kneeling	1½
Table III, 200 yards, standing (simulated)	½

WEDNESDAY

Table III, 200 yards, standing (simulated)	½
Table III, 200 yards, standing	1½
Table IV, 200 yards, sitting from standing (simulated)	½
Table IV, 200 yards, sitting from standing	1
Table IV, 200 yards, kneeling from standing (simulated)	½
Table IV, 200 yards, kneeling from standing	1
Table IV, 300 yards, prone from standing (simulated)	½
Table IV, 300 yards, prone from standing	1
Table III (simulated)	1½

THURSDAY

Table III (simulated)	1
Table IV, 200 yards, sitting from standing (simulated)	½
Table IV, 200 yards, sitting from standing	1
Table IV, 200 yards, kneeling from standing (simulated)	½
Table IV, 200 yards, kneeling from standing	1
Table IV, 300 yards (simulated)	½
Table IV, 300 yards	1
Table V	2½

FRIDAY

	Hours
Table VI, 200 yards (simulated)	½
Table VI, 200 yards	1½
Table VI, 300 yards (simulated)	½
Table VI, 300 yards	1½
Table VI, 500 yards (simulated)	½
Table VI, 500 yards	1½
Table VII and VIII (simulated)	2

SATURDAY

Table VII (simulated)	½
Table VII	2
Table VIII (simulated)	½
Table VIII	1

b. Courses B, C, and D.—The preparatory exercises are the same as course A. All other firing is conducted in a manner similar to course A with the time reduced accordingly.

■ 159. LECTURES AND DEMONSTRATIONS.—*a.* The lectures at the beginning of each step are an important part of the instruction. The lectures may be given to a group as large as a regiment or to a body of recruits of similar size. However, when a battalion takes up rifle training the talks and demonstrations usually are made by the captain or a lieutenant of each company. It is not necessary that they be expert shots.

b. The notes on lectures which follow are to be used merely as a guide. The points which experience has shown to be the ones that usually require elucidation and demonstration are placed in italic side headings. The notes which follow each heading are merely to assist the instructor in preparing his lecture. The lecturer should know in advance what he is going to say on the subject. In no circumstances will he read to a class the lecture outlines contained in this manual nor will he read a lecture prepared by himself. During the lecture, the headings in italic, jotted down as notes, serve as a guide to the order in which the points are to be discussed. If the instructor cannot talk interestingly and instructively on each subject without elaborate notes, he should not give the lectures at all.

c. It is important to explain and demonstrate to the men undergoing instruction just how to go through the exercises and to impress upon them why they are given.

■ 160. FIRST LECTURE: SIGHTING AND AIMING.—*a*. The class is assembled in a building or natural amphitheater in the open where all can hear the instructor and see the demonstrations.

b. The following equipment is necessary for the demonstrations:

(1) Adjustable sights and bull's-eye (made of cardboard).

(2) One sighting bar.

(3) One rifle rest.

(4) One rifle.

(5) One small sighting disk.

(6) One 10-inch sighting disk.

(7) One small box.

(8) Material for blackening sights.

c. The following subjects are usually discussed in the first lecture:

(1) *Value of knowing how to shoot*.—(*a*) The rifle is the principal weapon of the Infantry in war. Expertness in its use gives the individual confidence and a higher morale.

(*b*) Individual proficiency increases the efficiency of the unit as a whole.

(*c*) Rifle firing is good sport.

(2) *Object of target practice*.—(*a*) To teach men how to shoot.

(*b*) To show them how to teach others.

(*c*) To train future instructors.

(3) *Training to shoot well*.—(*a*) Any man can be taught to shoot well. Shooting is a purely mechanical operation which can be taught to anyone physically fit to be a soldier. It requires no inborn talent.

(*b*) There are only a few simple things to do in order to shoot well, but these things must be done exactly right. If they are done only approximately right, the results will be poor.

(4) *Method of instruction.*—(*a*) The method of instruction is the same as that used in teaching any mechanical operation.

(*b*) The instruction is divided into steps. The man is taught each step and practices it before going to the next step. When he has been taught all of the steps he is taken to the rifle range to apply what he has learned.

(*c*) If he has been properly taught the various preparatory steps, he will do good shooting from the very beginning of range practice.

(*d*) Explain coach-and-pupil method; why used.

(5) *Reflecting attitude of instructor.*—If the instructor is interested, enthusiastic, and energetic, the men will be the same. If the instructor is inattentive, careless, and bored, the men will be the same, and the scores will be low.

(6) *Examination of men on preparatory work.*—Each man is examined in the preparatory work before going to the range. An outline of this examination is given in paragraph 49.

(7) *Method of marking blank form.*—Explain blank form, paragraph 42*f*. Explain marking system by the use of a blackboard, if available.

(8) *Five essentials to good shooting.*—(*a*) Correct sighting and aiming.

(*b*) Correct position.

(*c*) Correct trigger squeeze.

(*d*) Correct application of rapid-fire principles.

(*e*) Knowledge of proper sight adjustment and aiming point adjustments.

(9) *Today's work.*—First step, sighting and aiming.

(10) *Demonstration of first sighting-and-aiming exercise.*—Have a squad on stage or platform show just how this exercise is carried on.

(11) *Blackening sights.*—Explain why this is done and demonstrate how.

(12) *Demonstration of second sighting-and-aiming exercise.*—Assume that some of the squad have qualified in the first exercise. Put these men through the second sighting-and-aiming exercise and show just how it is done.

(13) *Demonstration of third sighting-and-aiming exercise.*—(*a*) Assume that some of the squad have qualified in the second sighting-and-aiming exercise. Put these men through the third sighting-and-aiming exercise and show just how it is done.

(*b*) Show how the squad is organized for the coach-and-pupil method.

(14) *Long-range shot group work.*—Show the class the disk for 200-yard shot group work. Explain how this work is carried on and why. Show some simple system of signals that may be used.

(15) *Final word.*—(*a*) Start keeping your blank form today.

(*b*) Organize your work so that all men are busy at all times.

(16) Are there any questions?

(17) Next lecture will be ---------- (State hour and place.)

■ 161. SECOND LECTURE: POSITIONS.—*a*. The following equipment is necessary for the demonstrations in this lecture:

1 rifle with sling.
1 sandbag.
1 box with small aiming target.
1 aiming device.

b. The following subjects are usually discussed in the second lecture:

(1) *Importance of each step.*—(*a*) Each step includes all that has preceded.

(*b*) Each step must be thoroughly learned and practiced or the instruction will not be a success.

(2) *Necessity for correct positions.*—Good shots use the normal positions. Few men with poor positions are even fair shots. Few men with good positions are poor shots. Instruction in positions involves correct aiming and taking up the slack.

(3) *Gun sling.*—Demonstrate both of the gun-sling adjustments and explain why they are used and when each is used.

(4) *Taking up slack.*—Show the class the slack on the trigger. Explain why it is taken up in the position exercises. (Cannot begin to squeeze the trigger until the slack has been taken up.)

(5) *Holding breath.*—Explain the correct manner of holding the breath and have the class practice it a few times. Explain how the coach observes the pupil's breathing by watching his back.

(6) *Aiming device.*—Show how it is placed on the rifle and how it is used.

(7) *Position of thumb.*—Should be either over the stock or along side the stock. Explain why.

(8) *Joint of finger.*—Trigger may be pressed with first or second joint; second joint is preferable if it can be used without strain.

(9) *Prone position.*—(*a*) Demonstrate correct prone position, calling attention to the elements which go to make up a correct prone position: gun sling properly adjusted; body at the correct angle; legs spread well apart; position of the butt on the shoulder; position of the hands on the rifle; position of the cheek against the stock; position of the elbows.

(*b*) Mention the common errors which occur in the prone position.

(*c*) Demonstrate the correct position again.

(10) *Sandbag rest position.*—(*a*) Demonstrate in the same manner as described for the prone position.

(*b*) Demonstrate the coach adjusting the sandbag to the pupil.

(11) *Sitting position.*—Demonstrate in the same manner as described above for the prone position.

(12) *Kneeling position.*—Demonstrate in the same manner as described above for the prone position.

(13) *Standing position.*—Demonstrate in the same manner as described above for the prone position.

(14) *Today's work; position exercises.*—(*a*) Demonstrate the duties of the coach in a position exercise, calling attention to each item.

(*b*) Demonstrate the position of the coach. Always placed so that he can watch the pupil's finger and eye.

(*c*) Place a squad on an elevated platform and show how the squad leader organizes it by employing the coach-and-pupil method.

(*d*) Continue the long-range triangle work today.

(15) *Do not squeeze the trigger today.*—Take up the slack in these exercises but do not squeeze the trigger.

(16) *Keep the blank forms up to date.*—Examine each man in the squad at the end of the day's work and assign him a mark.

(17) Are there any questions?

(18) Next lecture will be ______________. (State hour and place.)

■ 162. THIRD LECTURE: TRIGGER SQUEEZE.—*a.* The following equipment is necessary for the demonstrations:

1 rifle with sling.
1 aiming device.
1 box with small aiming target.

b. The following subjects are usually discussed in the third lecture:

(1) *Trigger squeeze most important.*—Explain that there is only one correct method of squeezing the trigger—a steady increase of pressure so that the firer does not know when the explosion will take place. Emphasize the fact that this method of squeezing the trigger secures good results and must be used in rapid fire.

(2) *Machine rest example.*—Lay the rifle on the table, pointing down the room and toward an actual or imaginary target; assume that the rifle is in a machine rest which runs on a track parallel to the line of targets; assume that you fire a shot which hits the left edge of a 36-inch bull's-eye 1,000 yards away; then move the rifle 36 inches to the right on the table as if it were sliding along the parallel track and assume that another shot is fired. Where does it hit? Answer: The right edge of the bull's-eye. Move the rifle backward and forward between these two positions, and assume a shot is fired any time while it is moving. Where will it hit? Answer: In the bull's-eye. Now assume that you hold the butt of the rifle still and move the muzzle a fraction of an inch. Where will it hit? Answer: It will miss the whole target. It hits the target when the whole rifle moves, but misses it when only one end moves.

(3) *Pulsations of body.*—The natural movements of the body and its pulsations produce more or less parallel movement of the rifle. Often men who are apparently very unsteady make good scores. Thus you can see that if you squeeze the trigger so that you do not know when the rifle is going

off, the shot is displaced only by the amount of the parallel movement and will be a good one. But if you give the trigger a sudden jerk you deflect one end of the rifle and the shot will be a poor one.

(4) *Aim and hold.*—Any man can easily learn to hold a good aim for a long time—15 to 20 seconds. Poor shots are usually the men who spoil their aim when they fire.

(5) *Coach squeezing trigger.*—(*a*) The fact that when the coach squeezes the trigger for the firer the shot is almost invariably a good one proves that poor shooting is principally caused by errors in the trigger squeeze.

(*b*) It is not necessary for the coach to watch the sights through the aiming device. By watching the firer's back he knows when the firer is aiming and then presses steadily on the trigger. Demonstrate how it is done.

(6) *Flinchers.*—Flinching is an instinctive and subconscious act which no man can control. Any man who gives the trigger a sudden jerk will flinch. Good shots are the men who avoid this flinching by squeezing the trigger in such a way as not to know when the explosion will take place.

(7) *When rifle goes off before man is ready.*—Often a man who has been doing poor shooting will state upon firing a shot, "I cannot call that shot. It went off before I was ready." Almost invariably these shots are well placed. His poor shooting has been caused by *getting ready* for them.

(8) *Calling shot.*—Explain calling the shot and why it is done.

(9) *Today's work: trigger-squeeze exercise.*—(*a*) Demonstrate the duties of a coach in a trigger-squeeze exercise and call attention to each item.

(*b*) The work is carried on as in position exercises with the squeezing of the trigger added.

(*c*) Practice in the prone position only this morning, first with, then without, the sandbag.

(*d*) Finish up the long-range shot group work today.

(10) *Keep blank forms up to date.*—Examine each man in the squad at the end of the day's work and assign him a mark.

(11) *Final word.*—Do not let yourselves become bored with this work. It is easy to learn but it takes a lot of practice

to train the muscles and to get in the habit of doing the right thing without thinking.

(12) Are there any questions?

(13) Next lecture will be ------ (State hour and place.)

■ 163. FOURTH LECTURE: RAPID FIRE.—*a*. The following equipment is necessary for the demonstrations:

1 rifle with sling.

2 clips of corrugated practice dummy cartridges.

b. The following subjects are usually discussed in the fourth lecture:

(1) *Rapid fire—true test of good shot.*—Superiority of fire in battle depends on the ability to deliver rapid and accurate fire.

(2) *Fire to be accurate.*—(*a*) Fire that is rapid without being accurate is of no value.

(*b*) Trigger sqeezed the same as in slow fire.

(3) *Meaning of rapid fire.*—Rapid fire is merely continuous fire. The rapidity comes from working the bolt quickly, reloading the clips into the magazine smoothly, and keeping the eye on the target while operating the bolt.

(4) *Keeping eye on target.*—(*a*) Explain the advantages of this and how it gains time.

(*b*) Demonstrate the correct way to operate the bolt and the way a man does it who looks into the magazine each time.

(5) *Application in war.*—Explain the advantage of keeping the eye on the target in war.

(6) *Bolt-operation exercise.*—(*a*) Show how the follower is held down by the use of an empty clip.

(*b*) Demonstrate the exercise.

(7) *Operating bolt in rapid fire.*—Show how it is done in the prone position and in the sitting positions, calling attention to the details in each case.

(8) *Necessity for great amount of rapid-fire practice.*—(*a*) A smooth and rapid bolt operation on the part of a soldier materially increases his rapid-fire scores and his efficiency in battle.

(*b*) Practice in loading clips of cartridges into the magazine is also necessary.

(9) *Assuming positions rapidly.*—(*a*) The prone position can be assumed more quickly than any other and an aimed shot fired more rapidly from this position.

(*b*) Application in war.

(*c*) Demonstrate by the numbers and then as one smooth movement.

(*d*) Even if it takes a few seconds longer, get into the correct position before starting to shoot.

(10) *Today's work: rapid-fire exercise.*—(*a*) Explain how exercises are to be carried on.

(*b*) Demonstrate the duties of a coach in a rapid-fire exercise and call attention to each item.

(*c*) First period today will be devoted to the bolt-operation exercise and it will be repeated in short periods from time to time until each man qualifies.

(*d*) You will also practice taking positions rapidly.

(11) *Keep blank forms up to date.*—Examine each man in the squad at the end of the day's work and assign him a mark.

(12) Are there any questions?

(13) Next lecture will be ____________________ (State hour and place.)

■ 164. FIFTH LECTURE: EFFECT OF WIND AND LIGHT; SIGHT CHANGES; SCORE BOOK.—*a*. This part of the preparatory instruction can be given on any day in which the weather forces the work to be done indoors. If no bad weather occurs, this work should follow rapid-fire instruction.

b. The following equipment is necessary for the demonstrations:

(1) One A, B, and D target for each range at which each of these targets is to be used in range practice. These targets to be mounted on frames and marked with the proper windage and elevation lines.

(2) Ten spotters than can readily be stuck into the targets.

(3) Each man to have his rifle and a score book.

c. The following subjects are usually discussed in the fifth lecture:

(1) *Targets.*—(*a*) Explain the divisions on the targets and give the dimensions of each.

(*b*) Call attention to windage and elevation lines. Have class compare them with diagram in the score book. Explain why lines are farther apart as the range increases.

(2) *Weather conditions.*—All weather conditions are disregarded except wind.

(3) *Wind.*—(*a*) Explain how the direction of the wind is described.

(*b*) Explain how the velocity of the wind is estimated.

(*c*) Explain the effect of wind. Effect increases with distance from target.

(4) *Wind rule.*—State rule and explain it.

(5) *Elevation rule.*—State rule and explain it.

(6) *Light.*—Explain effect.

(7) *Shooting up or down hill.*—(*a*) Explain the effect on elevation.

(*b*) Remember this rule when shooting at hostile airplane.

(8) *Score book.*—(*a*) Explain the use of score book on range.

(*b*) Have class open score books and explain items of keeping a score point by point.

(9) *Exercises.*—Give the class a number of small problems like those in paragraph 48*h* to demonstrate how the day's work is to be carried on.

(10) *Today's work.*—(*a*) Study and practice in sight setting, sight changing, aiming point, and the use of score book. Squad leaders and other instructors will work up problems for their groups. Coach-and-pupil method is also used in which the coach states the conditions for the pupil.

(*b*) Additional practice in the exercises of the preceding days and rapid-fire exercises.

(11) Are there any questions?

(12) Next lecture will be______________. (State hour and place.)

■ 165. SIXTH LECTURE: RANGE PRACTICE.—This lecture and demonstration should immediately precede range firing. If the class is not too large, it should be given on a firing point of the rifle range.

a. The following equipment is necessary for the demonstrations:

(1) Material for blackening sight.

(2) One rifle with gun sling.

(3) One aiming device.

(4) Range dummy cartridges or practice dummy cartridges.

b. The following subjects are usually discussed in the sixth lecture:

(1) *Preparatory work applied.*—Range practice is carried on practically the same as a trigger-squeeze exercise except that ball cartridges are used.

(2) *Coaching.*—Coach watches the man not the target. Coach does not keep the score for the pupil. Pupil must make his own entries in his score book. Coach sees that he does this.

(3) *Officers and noncommissioned officers.*—(*a*) Supervise and prompt the men acting as coaches.

(*b*) Personally coach pupils who are having difficulty in making good scores.

(4) *Spotters.*—(*a*) Use in both slow and rapid fire.

(*b*) If a spotter near the edge of the bull's-eye bothers the pupil in aiming, it may be removed before he fires again.

(5) *Watching the eye.*—Explain how this indicates whether or not the pupil is squeezing the trigger properly.

(6) *Position of coach.*—Demonstrate in each one of the positions.

(7) *Demonstration of coaching in slow fire.*—(*a*) Place a man on the firing point and show just what a coach does by calling attention to each item. (See par. 56*e*(7).)

(*b*) Demonstrate the use of the aiming device.

(*c*) Demonstrate the use of dummy cartridges in slow fire.

(*d*) Demonstrate coach squeezing the trigger for pupil.

(8) *Demonstration of coaching in rapid fire.*—Same procedure as in paragraph 516*e*(8).

(9) *Use of range dummy cartridges in rapid fire.*—(*a*) Show how dummy cartridges are mixed with service cartridges for rapid-fire training and explain why this is done.

(*b*) Coaches must be alert in this kind of practice in order to prevent pupil from looking into the chamber.

(10) *Read final precautions for slow fire.*—See paragraph 56*f*.

SECTION IV

MARKSMANSHIP—AIR TARGETS

■ 166. PRELIMINARY PREPARATION.—*a.* The officer in charge of rifle antiaircraft training should be thoroughly familiar with the subject; should have sufficient officers detailed as assistant instructors; and should train the assistant instructors and a demonstration group before the first training period.

b. He should inspect the range and equipment far enough ahead of the first training period to permit correction of deficiencies.

■ 167. DESCRIPTION OF MINIATURE RANGE.—*a. Targets.*—(1) *Horizontal.*—This target is designed to represent a sleeve target towed by an airplane flying parallel to the firing point.

(2) *Double diving and climbing.*—This target is in two sections. The right section is designed to represent a sleeve target towed so as to pass obliquely across the front of the firing line in the manner of an airplane diving, if run from left to right, or climbing, if run from right to left. The left section is the same but represents an airplane diving from right to left and climbing from left to right.

(3) *Overhead.*—This target is designed to represent a sleeve target towed by an airplane which is approaching the firing line and will pass overhead, or when run in the opposite direction represents an airplane that has passed over the firing line.

b. Size and speed of silhouette.—The black silhouette is a representation at 500 inches of a 15-foot sleeve at a range of 330 yards. It is 7.5 inches long. The speed of the silhouette should be between 15 and 20 feet per second. This speed represents that of an airplane flying between 150 and 200 miles per hour at a range of 200 yards. The size and speed of the silhouette are based upon the time of flight of the caliber .22 bullet for 500 inches. This time of flight is approximately 0.04 second. When the target is moving at a speed of 15 feet or 180 inches per second it will move 180 times .04 or 7.2 inches. Therefore in order to hit the silhouette the aim must be directed approximately one silhouette length in front of it. If two or three target-length (silhouette-length) leads are used, the shot will hit in the appropriate scoring spaces. This does not hold equally true on the overhead target. If the shot is fired when the range is less than 500 inches from the firer the lead necessary will be less than one target length.

■ 168. PREPARATORY EXERCISES.—*a.* A method of conducting the preparatory exercises is given in paragraph 83.

b. Each assistant instructor is assigned a target and conducts the preparatory training and firing of all groups on his target.

c. In preparatory training the coach and pupil should change places frequently.

d. Forty-five minutes at each type of target should be sufficient to train each soldier in the preparatory exercises.

e. A detail of one noncommissioned officer and four to six men should be provided to operate each type of target.

■ 169. MINIATURE RANGE FIRING.—*a. Caliber .22 rifle.*—(1) The rifle should have the open sight.

(2) Two magazines for each caliber .22 rifle should be provided.

(3) Ammunition should be available immediately in rear of the firing line at each type of target.

(4) Coaches should load magazines as they become empty.

(5) Scorers should be detailed for each type of target. After each score is fired, they score the target. They call off the number of hits made on each silhouette and pencil the shot holes. The coaches enter the scores on the firer's score card.

(6) A platform permitting the scorer to score the target should be provided for each type of target.

(7) To stimulate interest, the instruction can be concluded with a competition between individuals, squads, or training groups.

(8) If available, targets as shown in figure 57 may be used on nonoverhead targets for group firing or competitions. Only one target-length lead may be used in firing on this target.

(9) Considerable supervision is required in order to maintain target operation at the proper speed. This speed is necessary because the lead is based upon a speed of 15 to 20 feet per second.

FIGURE 57.—Antiaircraft targets.

(10) Safety precautions must be constantly observed.

b. Caliber .30 rifle.—If the size of the danger area permits, the caliber .30 rifle may be fired on the miniature range. This firing is conducted in the same manner as with the caliber .22 rifle with the following exceptions:

(1) The top of the battle peep sight is used (see par. 81*c*).

(2) The lead necessary to hit the black silhouette is about 2.5 inches. This reduction is caused by the difference in the time of flight of the caliber .30 and caliber .22 bullets for

500 inches. The time of flight of the caliber .30 bullet for 500 inches is 0.015 second. When the target is operated at the speed of 15 or 20 feet per second the silhouette will move approximately 2.5 inches during the time of flight of the bullet.

(3) Using the top of the battle peep sight, the line of aim is lower than the trajectory of the bullet. Therefore it is necessary to aim low in order to hit the silhouette. The aim should be directed at a point about 4 inches below the line of travel of the silhouette.

(4) Men must be constantly cautioned to keep the weight of the body forward. This is to prevent them from being pushed over by the recoil of the weapon.

(5) Preparatory exercises using the caliber .30 rifle should precede firing with that weapon. These exercises are conducted as explained for the caliber .22 rifle.

(6) The interval between individuals on the firing line should be increased. This may be accomplished by placing only one-half the group on the firing line at one time.

■ 170. TOWED-TARGET FIRING.—*a. Range organization.*—(1) All firing at towed targets is done by a unit of such size that its fire can be readily controlled and directed. The platoon is the most convenient unit for such firing.

(2) An *ammunition line* should be established 10 yards in rear of the firing line. Small tables at the rate of one per 10 men in a firing group are desirable.

(3) Immediately in rear of the ammunition line the *ready line* should be established.

(4) The first platoon or group to fire is deployed along the ready line with individuals in rear of their places on the firing line. Other groups are similarly deployed in a series of lines in rear of the first.

(5) Upon command of the officer in charge, the group on the ready line moves forward to the firing line securing ammunition en route; other groups close up.

(6) As each group completes its firing it moves off the firing line, passing around the flanks of the ready line so as not to interfere with the group moving forward.

(7) An ammunition detail is provided to issue ammunition

to groups as they move forward to the firing line and to collect unfired ammunition from each group as it completes firing. These two operations should be performed simultaneously. Unfired ammunition is delivered to the statistical officer.

(8) The officer in charge should have at least 3 assistants—2 safety officers and 1 statistical officer.

b. Ammunition.—(1) Ball or tracer ammunition may be be used. Tracer ammunition is of no assistance to the firer; it is useful to show the groups waiting to fire the size and density of the cone of fire delivered by the firing group.

(2) Tracer ammunition assists the officer in charge in verifying the lead announced in the fire order. It also provides a means of checking the firer's estimate of the lead ordered.

c. Fire distribution.—(1) *Leads.*—(*a*) The maximum lead necessary to engage a sleeve target while it is within the firing area is required when it first enters the area and as it leaves the area. The minimum lead necessary is required when the sleeve is directly opposite the firing line. The lead used in the normal method of fire distribution is the average of these two extremes. For example, if the maximum slant range is 600 yards and the minimum slant range is 300 yards, the lead used would be that required for a slant range of 450 yards. All fire is delivered with the same lead.

(*b*) The lead table given below may be helpful. It is based upon a 15-foot sleeve towed at 200 miles per hour and M–2 ammunition.

Slant range	Lead required
100	2
200	5
300	8
400	11
500	14
600	18

The above leads were computed using the formula $\frac{V \times T}{L}$ equals target-length leads required. This simple formula gives sufficiently accurate results for all practical purposes. V equals speed of target—200 mph or approximately 300 feet per second. T equals time of flight of bullet. L equals length of target—15 feet. For a slant range of 300 yards: $\frac{300 \times .38}{15} = 7.6$ or 8 target-length leads. If the speed of the target is not 200 mph the lead to be used may be easily determined. For example, speed of target is 140 mph; 140 is seven-tenths of 200. Therefore at 300 yards slant range the lead necessary is $0.7 \times 8 = 6$ target-length leads.

(2) *Methods.*—The normal method of fire distribution is given in chapter 4. This method will be taught in towed-target range practice. If time and ammunition allowances permit, other methods may also be taught.

(3) *Variable lead method.*—(*a*) Using this method the individual rifleman fires each shot with a different lead. The maximum lead is used as the target enters and leaves the firing area. The minimum lead is used when the target is directly opposite the firing line. For example, three rounds are to be fired as the sleeve target passes across the front of the firing line. The first round is fired shortly after the target enters the firing area; the second round is fired when the target is near the center of the firing area; the third shot is fired shortly before the sleeve leaves the firing area. The fire order given by the officer in charge is: 1. SLEEVE TARGET APPROACHING FROM THE LEFT (RIGHT), 2. 3 ROUNDS LOAD, 3. 14–8–14 TARGET-LENGTH LEADS, 4. COMMENCE FIRING. In this example it is expected that the three shots will be fired at slant ranges of approximately 500 yards, 300 yards, and 500 yards, respectively.

(*b*) This method has given good results but is more difficult to apply than the normal method.

d. Safety precautions.—Safety precautions as given in paragraph 93 must be strictly enforced. This requires constant supervision by the officer in charge.

e. Results.—The results of all towed-target firing should be recorded and analyzed. The statistical officer should record the total number of rounds fired and the hits obtained on each target. If the number of hits falls below the number expected, the reason should be sought and explained to the men. On the other hand when results are satisfactory the men should be impressed with the value of rifle antiaircraft fire. The results of firing at sleeve targets moving at speeds of 130 to 150 miles per hour indicate that about one percent hits can be expected.

Section V

TECHNIQUE OF FIRE

■ 171. General.—The instructor should secure the necessary equipment, inspect ranges, and detail and train the necessary assistants, including demonstration units, prior to the first period of instruction. Instructors should use their initiative in arranging additional exercises in the application of the principles involved. It should be explained to trainees how the exercises used illustrate the various principles in the technique of fire. Good work in the execution of the exercises as well as errors should be called to the attention of all trainees.

■ 172. Range Estimation.—*a.* A number of ranges to prominent points on the terrain should be measured so that a few minutes of each period can be devoted to range estimation.

b. Range cards as shown below are helpful in figuring percentage of errors.

RANGE ESTIMATION

Name..........

Company..........

Squad..........

Number	Estimate	Correct	Per-cent	Remarks	Number	Estimate	Correct	Per-cent	Remarks
1					21				
2					22				
3					23				
4					24				
5					25				
6					26				
7					27				
8					28				
9					29				
10					30				
11					31				
12					32				
13					33				
14					34				
15					35				
16					36				
17					37				
18					38				
19					39				
20					40				

(Front)

TABLE FOR COMPUTING ERRORS IN RANGE ESTIMATION

Range, in yards	Error, in yards										
	5	10	15	20	25	30	35	40	45	50	100
250	2	4	6	8	10	12	14	16	18	20	40
275	2	4	5	8	9	11	13	15	16	18	36
300	2	3	5	7	8	10	12	13	15	17	33
330	2	3	5	6	8	9	11	12	14	15	30
350	1	3	4	6	7	9	10	11	13	14	29
380	1	3	4	5	7	8	9	11	12	13	26
400	1	3	4	5	6	8	9	10	11	13	25
420	1	2	4	5	6	7	8	10	11	12	24
440	1	2	3	4	6	7	8	9	10	11	23
460	1	2	3	4	5	7	8	9	10	11	22
480	1	2	3	4	5	6	7	8	9	10	21
500	1	2	3	4	5	6	7	8	9	10	20
520	1	2	3	4	5	6	7	8	9	10	19
540	1	2	3	4	5	6	7	8	9	10	19
560	1	2	3	4	4	5	6	7	8	9	18
580	1	2	3	3	4	5	6	7	8	9	17
600	1	2	3	3	4	5	6	7	8	8	17
620	1	2	2	3	4	5	5	6	7	8	16
640	1	2	2	3	4	5	5	6	7	8	16
660	1	2	2	3	4	5	5	6	7	8	15
680	1	1	2	3	4	4	5	6	7	8	15
700	1	1	2	3	3	4	5	6	6	7	14
720	1	1	2	3	3	4	5	6	6	7	14
740	1	1	2	3	3	4	5	6	6	7	14
760	0	1	2	3	3	4	5	5	6	7	13
780	0	1	2	3	3	4	4	5	6	6	13
800	0	1	2	3	3	4	4	5	6	6	13
850	0	1	2	2	3	3	4	5	5	6	12
900	0	1	2	2	3	3	4	4	5	6	11
950	0	1	2	2	3	3	4	4	5	5	10
1,000	0	1	2	2	3	3	4	4	5	5	11

NOTE.—Example of the use of this table: Suppose the correct range to be 695 yards and the estimated range to be 635. The "error in estimate" is consequently 60 yards. Select two "errors in estimate" in the 700-yard space (the nearest to the correct range given in the table) whose sum is 60 yards, as 50 and 10. Add the percentages shown thereunder, and the result will be approximately your error. In this case:

$$7+1=8 \text{ percent.}$$

(Back)

■ 173. TARGET DESIGNATION.—The major part of the time devoted to target designation should be spent on oral description. Simple designations should be required at first. This instruction should not be confined to the landscape panels.

■ 174. RIFLE FIRE AND ITS EFFECT.—This step of instruction can best be covered by explaining it on the blackboard, then by having several riflemen fire tracer bullets to demonstrate the trajectory, danger space, dispersion and classes of fire.

■ 175. APPLICATION OF FIRE.—*a*. Sufficient time and explanation should be devoted to the method of fire distribution to insure that all men fully understand it and can explain it in their own words.

b. A demonstration squad simulating fire should suffice to show the technique employed in assault fire.

■ 176. LANDSCAPE-TARGET FIRING.—*a*. An explanation and demonstration are necessary to show the technique and procedure of zeroing rifles and the firing of exercises on the landscape targets.

b. Units should be given practical work in writing fire orders for targets on the landscape panels prior to firing any exercises.

■ 177. FIRING AT FIELD TARGETS.—*a*. The most difficult factor in the preparation of problems for field firing is the selection of the terrain which complies with the safety regulations contained in AR 750–10. A drawing should be made on a map showing all safety angles, target positions, etc.

b. The appearance of the ordinary prone or kneeling silhouette depends a great deal upon the direction of the sun, the background, and the angle at which the targets are placed. The effect of solidity can be obtained by using two figures placed at right angles to each other. The effect of fire distribution on a linear target can be determined by using a screen of E targets nailed end to end; the screen should be located so as not to disclose the position of concealed targets.

c. As many squads should fire from one position at one time as the terrain permits. This firing should be controlled from a central location. Telephone communication between the firing point and the pits facilitates this instruction. During this training, individuals and units should approach and occupy their firing positions with due regard to cover and concealment, after which men are rearranged on the firing position according to the requirements of safety.

d. When sufficient time and ammunition are available, platoon exercises should be conducted.

e. About 60 to 70 percent of the score allotted for the grading of units should be given for such parts of the exercise as the approach march and occupation of the firing position, fire orders, time required to open fire, rate of fire, and fire control. The remaining 30 or 40 percent should be given for the number of hits on the targets and the number of targets hit.

INDEX

INDEX

INDEX

ISBN#978-1-940453-14-9
www.PeriscopeFilm.com